THE
SECURITY+ EXAM
GUIDE

TestTaker's Guide Series

THE SECURITY+ EXAM GUIDE

TestTaker's Guide Series

Christopher A: Crayton

CHARLES RIVER MEDIA, INC.
Hingham, Massachusetts

Publisher: David Pallai
Production: Publishers' Design and Production Services, Inc.
Cover Design: The Printed Image

CHARLES RIVER MEDIA, INC.
10 Downer Avenue
Hingham, Massachusetts 02043
781-740-0400
781-740-8816 (FAX)
info@charlesriver.com
www.charlesriver.com

This book is printed on acid-free paper.

Christopher A. Crayton. *The Security+ Exam Guide: TestTaker's Guide Series.*
ISBN: 1-58450-251-7

All brand names and product names mentioned in this book are trademarks or service marks of their respective companies. Any omission or misuse (of any kind) of service marks or trademarks should not be regarded as intent to infringe on the property of others. The publisher recognizes and respects all marks used by companies, manufacturers, and developers as a means to distinguish their products.

Library of Congress Cataloging-in-Publication Data

Crayton, Christopher A.
    Security+ exam guide / Christopher A. Crayton.
        p.   cm.—(TestTaker's guide series)
    ISBN 1-58450-251-7 — ISBN 1-58450-251-7
        1. Computer networks—Security measures—Examdinations—Study guides.
    2. Computer networks—Security measures—Examinations, questions, etc.
    3. Telecommunications engineers—Certification.   I. Title.   II. Series.
    TK5105.59.C73 2003
    005.8—dc21
                            2002154762

Printed in the United States of America
03 7 6 5 4 3 2 First Edition

CHARLES RIVER MEDIA titles are available for site license or bulk purchase by institutions, user groups, corporations, etc. For additional information, please contact the Special Sales Department at 781-740-0400.

Requests for replacement of a defective CD-ROM must be accompanied by the original disc, your mailing address, telephone number, date of purchase and purchase price. Please state the nature of the problem, and send the information to CHARLES RIVER MEDIA, INC., 10 Downer Avenue, Hingham, Massachusetts 02043. CRM's sole obligation to the purchaser is to replace the disc, based on defective materials or faulty workmanship, but not on the operation or functionality of the product.

This book is dedicated to all of my students, all certification exam test takers who have the courage to change their lives by achieving certification success, and those who spend countless hours defending information systems and network infrastructure from outside destructive influences.

CONTENTS

ACKNOWLEDGMENTS

First and foremost, I would like to thank David Pallai, president and owner of Charles River Media, for making this project a complete success. Once again, it has been an honor and a privilege to work with such a fine publishing company. Charles River Media's goal continues to mirror mine: provide the most useful certification test preparation books available.

Thank you to Darren Toback for his contribution to this book. Darren's security-related knowledge and experience have assisted in development of this powerful Security+ exam study resource.

ON THE CD

Thank you to coworker and ace Web developer Mark Porter for the design of the superb testing software package included with the CD-ROM that accompanies this book. Its design and ease of use displays the many talents of a great software developer.

Special thanks go to Posse Bob, Scott Gaertner, and Doug Morgan for protecting a very important corporate network infrastructure from external malicious influences.

Thanks to Jonathan Thatcher, Kris Madura, and CompTIA for involving me with Security+ Web conferences. The updates as to the development of the Security+ exam assisted with a structured approach in the creation of this fine study tool.

Thank you to Analin Alvarado, Michael Suhrbier, Julie Peirce, Brian King, Mr. Bear, Chris Gair, Micah Huston, Neil Strawbridge, Bill Burch, Shane Lewis, and Jim Morris for their assistance and support throughout this project. Without them, the creation of this book would not have been possible.

To Amanda Hutchinson, Nancy and Ken Crayton, Carol and David Brodie Howard, Amy Toback—thank you for your patience, support, and steadfast love.

Finally, thank you to my mentor and hero for the last two years, Steve Bleile. It has been an honor and a privilege to learn from someone of your caliber and nature. You are truly the most dedicated person I have ever met.

AUTHOR BIOGRAPHY

Christopher A. Crayton is a Certified A+/Network+ instructor, recognized as "Teacher of the Year" by Keiser College in 2000. Chris, a former mainframe operation specialist for Eastman Kodak's headquarters in Rochester, NY, has more than 20 years of experience in information security. He resides in Sarasota, Florida where he serves as Network Administrator for Protocol, an ECRM company. Christopher is a Microsoft Certified Systems Engineer (MCSE).

Darren M. Toback (contributor of Chapters 5 and 7 of this book and 25 of the test questions on the CD-ROM) has over 15 years of experience in various fields of computing, including Web programming and remote-office applications. His background also consists of securing medical billing systems for hospitals and directing on-the-job IT training programs. He is currently a freelance IT consultant in Sarasota, Florida where he lives with his wife, Amy. Darren is a Microsoft Certified Systems Engineer (MCSE).

CHAPTER 1

INTRODUCTION

IN THIS CHAPTER

- The CompTIA Security+ Certification Exam
- Registering for the Security+ Exam
- Test Site Requirements
- The Exam Structure
- Preparing Yourself for the Exam
- Useful Tools, Tips, and Study Techniques
- Book Structure and Sample Review Questions
- Chapter Summary
- References

Welcome and congratulations! You hold in your hands the most useful Security+ certification preparation study tool available. This book exists for one primary purpose: to help you achieve your goal of becoming CompTIA Security+ certified as soon as possible with one and only one resource—**this book**.

What separates this book from most of the other security exam and certification study books on the market? Unlike many other certification preparation study tools, the author and contributors of this book have real-life experience implementing and maintaining security processes and procedures as well as years of experience taking and passing certification exams. In other words, we are the real deal, we are not "paper certified specialists" authoring topics implemented by professional IT specialists.

It is also very important for you to know that the author of this book was in continuous contact with CompTIA regarding the Security+ exam development and objectives. This book has been crafted upon specific CompTIA information as it pertains to the development of the Security+ domain objectives. In other words, **this is the Security+ book to study if you want to get Security+ certified!**

In this chapter, you will learn the importance of this coveted certification, why it was developed, and how it can affect directly your present as well as future career goals in the information technology arena. You will be led through the entire CompTIA certification process from exam registration to sitting down and taking the exam. The exam structure will be explained in detail and you will receive advice from certification experts on how to prepare yourself both mentally and physically for the exam. Finally, you will be introduced to the book's structure and convinced of why this book should be your study tool of choice for the CompTIA Security+ examination.

This book was written as a no-nonsense, straight-to-the-point guide to prepare you to take and pass the CompTIA Security+ Exam. It is not intended as a replacement for hands-on training, nor will it prepare you to be an expert security specialist. It is not designed to fill your head with unneeded, useless information before you enter the exam site; that type of preparation is usually accompanied by failure. The Security+ exam is not filled with questions that ask you to refer to many useless pictures, charts, and graphs. For that reason, this book does not present you with many pictures or other graphics that are not exam specific. This book isolates particular topics that are most likely going to be addressed on the CompTIA Security+ exam.

If you are interested in gainful employment in the technologies industry, you should be certified. The majority of businesses today require applicants to provide proof of certification status. If you are applying for a position as a PC technician, network administrator, systems engineer, software engineer, auditor, security specialist, programmer, or developer, Security+ certification is for you.

THE COMPTIA SECURITY+ CERTIFICATION EXAM

Why did CompTIA create this certification and exam? The answer to this question is quite simple: the technologies industry as well as government agencies are screaming for qualified security specialists to implement and maintain security technologies and policies.

CompTIA has responded to the industry's high demand for skilled security specialists by creating this entry-level Security+ certification, which is also known as "The Foot Soldiers Certification." Its primary purpose is to verify that examinees possess the skills necessary to implement and support security policies and procedures. It should be noted that the Security+ exam is not vendor specific. In other words, CompTIA states that the exam does not focus on one operating system by any one vendor. All terms, concepts, principles, and theories can be applied to operating systems such as Windows and Linux. Keeping this in mind, one can most likely assume that the exam will simply focus on concepts.

WHY SHOULD YOU GET COMPTIA SECURITY+ CERTIFIED?

Passing the Security+ exam proves that you have the basic knowledge and skills necessary to implement and support security processes and procedures in general. If you are looking for a position in the technologies industry currently, obtaining this certification might just give you the edge you need to secure a position. If you are already employed, acquiring this certification might help you advance within your current position or increase your chances for future career growth within your company. The simple fact of the matter is this: anyone who has this certification title is going to be more valuable to the industry and have a better chance of securing employment than those who don't.

WHAT INFORMATION IS COVERED IN THE SECURITY+ EXAM?

CompTIA organizes the security information that will be covered in its certification exams into what are called *domains*. CompTIA has chosen the following five domains and sub-domain topics from which questions on the Security+ Exam will be drawn:

- General security concepts
- Communications security

- Infrastructure security
- Basics of cryptography
- Operational/organizational security

Each of these domains is comprised of various security-related topics that range from firewalls to digital certificates. Please refer to the Table of Contents of this book for more information regarding the specific topics covered for each of CompTIA's specific domains. This book covers all of the CompTIA Security+ domains. It is important to note that security is a huge subject. There are thousands of books on the market that are based completely on a single security related subject such as cryptography. This book will cover a bit more security-related information than those specified in the CompTIA Security+ domain objectives. However, it will not overburden you with useless non-security related material. This information is in place to give you a strong overview of security subject matter in general with a very fine microscope on the subject matter CompTIA is most likely to target with specific questions.

In order to prepare yourself for the exam as well as possible, you need to isolate the areas that will be targeted on the exam and focus your time and energy fine tuning your skills with these subjects.

General Security Concepts

This domain encompasses the following components:

- Access control
- Authentication
- Non-essential services and protocols
- Attacks
- Malicious code
- Social engineering
- Auditing

COMMUNICATIONS SECURITY

This domain encompasses the following components:

- Remote access
- E-mail
- Web
- Directory

- File transfer
- Wireless

INFRASTRUCTURE SECURITY

This domain encompasses the following components:

- Devices
- Media
- Security topologies
- Intrusion detection
- Security baselines

BASICS OF CRYPTOGRAPHY

This domain encompasses the following components:

- Algorithms
- Concepts of using cryptography
- PKI
- Standards and protocols
- Key management/certificate lifecycle

OPERATIONAL/ORGANIZATIONAL SECURITY

This domain encompasses the following components:

- Physical security
- Disaster recovery
- Business continuity
- Policy and procedures
- Privilege management
- Forensics
- Risk identification
- Education
- Documentation

WHAT ARE THE PREREQUISITES FOR THE EXAM?

CompTIA states that individuals wishing to take the Security+ exam should have A+/Network+ certification, or equivalent, and at least two years of hands-on networking and TCP/IP experience. It is also suggested that you have this experience before using this book as a study tool for the Security+ exam. The author of this book suggests specific hands-on experience with authentication mechanisms, backup and recovery tools, digital certificates, and business recovery and response implementation procedures.

REGISTERING FOR THE SECURITY+ EXAM

This book has you and your certification status in mind. You, the examinee, must reduce the stress levels involved with preparation and scheduling for the exams in order to focus your energy on your goal: getting Security+ certified.

In the United States and Canada, there are two companies that you can register with to take the Security+ certification exam as well as other certification exams: VUE and Prometric. You can register online or call either company. To register with VUE online, visit *http://www.vue.com/comptia* or call 1-877-551-PLUS (7587). To register with Prometric online, visit *http://www.2test.com* or call 1-800-909-3926. You can also register with either company and get worldwide registration information directly through CompTIA.

ON THE CD   VUE and Prometric both support different testing centers. Who you register with will depend on where you want to take the test.

You are required to register at least 12 hours before you take the test. If you decide to cancel after registering, you must call 12 hours before your scheduled test time to cancel or you will forfeit your money. Your best bet is to pick a target date for taking the exam and then give yourself at least 30 days to study this book. Make sure that you can answer all questions correctly that are included on the testing preparation CD-ROM that accompanies this book. Next, register at least three days ahead of the date and time you wish to sit for the exam. Ask the registration person at VUE or Prometric for the nearest test center location in your area.

Currently, the cost of the Security+ exam is approximately $225.00 US for non-CompTIA members. If you want to receive CompTIA certification exam membership prices and discounts for CompTIA certifications, you will need to become a CompTIA member. The current cost of the Security+ exam for CompTIA members is approximately $175.00 US dollars. Please see the CompTIA Web site for more details regarding membership.

You do not get your money back if you fail the exam. The exam is currently offered in English. If you pass the exam, you are instantly CompTIA Security+ certified. All CompTIA certifications are good for a lifetime. You will not have to recertify. A welcome kit will be sent to you within four to six weeks, depending on shipping and other factors. If you fail the exam, you can reregister, reschedule, and retake the exam at any time. You will have to pay again to retake the exam.

You need to have the following information ready before you register for the exam:

- **Contact information:** Your name, company name, mailing address, and e-mail address.
- **The exam name and number you wish to take:** In this case, Security+ Exam SY0-01.
- **Your method of payment:** Your best bet is to pay with a valid credit card for quick registration and get the testing date and time slot that fits your schedule. Other forms of payment must be received by the registration center before the tests can be scheduled.

TEST SITE REQUIREMENTS

It is advised that you arrive at the test site one hour before your exam. Take a little time to get comfortable with the surroundings at the test site, sign in at the registration desk, and take the time to study last-minute charts and details that you feel are your weak points. I suggest you use the Test Tips at the end of each chapter in this book as a final review preparation. Depending on the schedule of exams at the test site, generally, you can take care of the paperwork, find a quiet area to do some last-minute cramming, and then go take the test. If no one is scheduled before you, it might be possible to take the tests before your scheduled time.

You will be asked to provide two forms of identification at the testing site: a valid driver's license and a credit card are sufficient. One form of identification must be a photo identification. You might be asked for two forms of ID that display your signature. When you are ready to enter the testing room, you will be given one or two blank sheets of paper and a marker or pencil. You cannot take any other books or notes with you. Cell phones are not allowed and you will not be allowed to leave the examination room until you finish the exam. Once you have signed in at the testing center, a testing coordinator will direct you to a computer that will have your test ready to go. You might be required to enter a security ID before you start the test. This ID is normally your social security number. The testing coordinator will inform you of any special procedures for the particular test center.

The CompTIA certification program has been changed recently to produce a more valuable and desired certification. This book is published with up-to-date information in relation to the current Security+ exam. For more information about the CompTIA Security+ certification program and other CompTIA certifications, visit the CompTIA Web site: *www.comptia.org.*

THE EXAM STRUCTURE

Similar to other CompTIA certification exams, the Security+ exam has been released in traditional linear format as opposed to adaptive testing format, otherwise known as CAT (computerized adaptive testing). You will be required to answer 100 multiple-choice questions within 90 minutes.

When you have completed answering the required number of questions, your score will be calculated and the test results will be displayed on your screen. You will know instantly whether or not you have passed the exam. Proceed to the test coordinator's desk and get your printed test results.

Here is a typical breakdown of the Security+ exam design:

- **You will be allotted 90 minutes to complete the exam:** You will see a timer or clock in the upper right-hand corner of the test screen that displays the amount of time you have remaining to complete the test.

- **You will be required to answer 100 multiple-choice questions within 90 minutes:** The minimum passing score is 764, on a graded scale of 100-900. Be careful—if CompTIA chooses, some questions might require more than one answer. If there are circles next to your choices, you must choose only one answer. However, if you see squares next to your choices, you will have the option to select one or more answers. Read the questions carefully. They usually say, "Choose two" or "Choose three." If a question asks you to "Choose two" and you select only one answer, the test will prompt you to choose two before you can proceed. The same is true for "Choose three" and so on. It is most likely that your Security+ exam will have only circles and require only one answer to each question. Keep an eye out for questions that offer you four "circle" choices where the last choice is "None of the above." These questions are tricky and deserve your special attention.

- **There might be questions that require you to click on a graphic radio button to display a diagram or image:** You'll be asked to select the correct answer or best choice from the displayed diagram or image. Some diagrams require you to select or identify several choices. You will be allowed to refer back to the diagram to make your selections. This is a good place to use your scratch paper or plastic card given to you by the exam proctor to keep your thoughts straight.

- **As is the case with CompTIA's traditional exam structure, you will have the option to mark a question for later review:** There will be a radio button on the screen identified with the word, *mark*, or "mark for later review." Simply place a check mark in the box. When you have finished the last question on the exam, you will have the option to go back through the questions you have marked and change your answers if you so choose. This is a tool used to remind you of what questions you had difficulty with or simply need to review before you end the exam.

PREPARING YOURSELF FOR THE EXAM

ON THE CD

It is important to focus when studying for and finally taking a certification exam. Make sure that you give yourself time daily to study this book and its accompanying practice test CD-ROM without any distraction. It is not a good practice to study when you are tired and unable to retain the required information. You want to be well rested and in a good frame of mind when you take this test. Don't stay up all night before taking your exam trying to cram 15 years of technical security related information into your head.

Confidence, in hand with good study habits, plays a very big role in this process and increases your chances of success. When preparing to take a certification exam, not only prepare to pass, but prepare to score as close to 100 percent as possible.

Make sure this is a positive experience to set the stage for your future certification testing goals. Learn to develop good study habits early on in your certification career. This book is based on information, tools, and techniques that have springboarded thousands of students and professionals to certification success and beyond.

USEFUL TOOLS, TIPS, AND STUDY TECHNIQUES

It seems that most certification preparation guides are geared to have you figure out some sort of magical strategy to answer questions correctly and ultimately pass certification exams. From firsthand experience, it's really quite simple. Prepare yourself well with proper study and choose the right answer to each question on the exam.

The book you are holding in your hands is the key to your Security+ certification success. Read the entire book twice. It has been crafted with your certification success in mind.

ON THE CD
The practice tests included on the CD-ROM are very accurate and are good simulations of the real tests. Take the practice tests repeatedly until you score 100 percent every time. When you take a test, whether it is a practice test or a real test, read each question carefully and go with your initial choice. Try not to read too much into the questions. Certification developers are great at making the wrong answers look good. They include key words in the questions that are used to confuse you. Words or phrases such as "Choose the best answer or possible solution," or "Most likely," or "Which best describes" will "Most likely" appear on this exam. The longer you sit there staring at these phrases on the screen, the more likely it is that you will pick the wrong answer. Focus completely on the questions in front of you and try not to think of anything else. If you have prepared for the exam properly, you will most likely know the correct answers to the questions on this entry-level examination. Your challenge will be to figure out what it is they are actually asking. The trick to mastering this exam lies in the question being asked, not the answers!

Learn to prepare yourself well and always remember the difference between pass and fail can be one question.

Test takers who have gone before you post *brain dumps* on many Internet sites. These dumps are usually questions that they remember seeing on their tests. Be very careful if you study these postings. These brain dumps might be helpful on the test but you should know the material thoroughly. Besides, many of these dumps contain incorrect answers and information.

Here's another very useful tip: it is the author's opinion that the following Internet site is the very best security resource available as a tool to keep you current with security trends and technical security related information—*http://searchsecurity.techtarget.com*

If you study this book, take the suggestions, and do the groundwork, the correct answers might just jump out as you take the real tests.

BOOK STRUCTURE AND SAMPLE REVIEW QUESTIONS

This book is part of the Charles River Media *TestTaker's Guide* series. It has been crafted and structured in a fashion that completely has your certification objectives in mind. The structure is based on a proven study method that will increase your chances of getting certified. It is completely up to you to utilize the tools at your fingertips to increase your own odds of passing the exam.

The chapters in this book are based on the specified CompTIA domains as they relate to Security+ study. Each chapter in this book includes a section called "Test Tips." The Test Tips are located towards the end of each chapter. These Test Tips are included to highlight some of the important items that might be targeted on the exam. At the end of each chapter in this book, you will also be able to test your knowledge by answering several chapter content related review questions. You might see questions relative to earlier chapters as you move forward. This design is in place to help you collectively build upon your skill sets as you move through the book. At the end of the book, you will find a TestTaker's exam that is available for you to test your cumulative knowledge of the combined chapters.

Use a piece of paper to cover the answers provided underneath the questions; don't cheat. If you do not understand the question or the answer, you might have to go back and do some review work. Pay very close attention! You might see questions regarding similar content on the actual exams.

SAMPLE REVIEW QUESTIONS

(Notice the circles next to your choices. You will select one answer.)

1. **Of the following choices, which item does not represent physical security?**
   - ○ A. Padlock.
   - ○ B. Gate.
   - ○ C. Security personnel.
   - ○ D. Encryption.

   Correct answer=D

2. **Which statement is true regarding firewalls?**
   - ○ A. They will protect your internal network from a virus that resides on a workstation on your internal network.
   - ○ B. If properly configured, a firewall will protect an internal network from an external network.
   - ○ C. Firewalls are used to protect your internal network from unauthorized external access through dial-up modem connections.
   - ○ D. Firewalls are used to protect internal server-room computers from external natural disasters.

   Correct answer=B

   (Notice the squares next to your choices. You will select one or more answers.)

3. **Which of the following are true statements regarding Kerberos tickets? (Choose three)**
   - □ A. They will authenticate users on the network and allow access to client computers.
   - □ B. They are used to allow users' access to objects.
   - □ C. A small yellow key appears in the corner of your window if you have an active Kerberos ticket.
   - □ D. If you use a service that requires a ticket, you will need to enter a network ID and password.

   Correct Answers=B, C, and D

CHAPTER SUMMARY

This chapter introduced you to the CompTIA Security+ certification exam. At this point, you should have a clear understanding of what the Security+ exam is about, how to register for the exam, the exam structure, and how to use this book to properly prepare yourself to pass this tough exam. As you begin your study of this subject, it is important to keep in mind that this exam was developed with entry- to mid-level certification in mind. However, fair warning! Do not take your study of this exam lightly and think you're going to pass this one with minimal effort. If you don't know the material, you will probably not pass. CompTIA has created this certification title and exam to meet the industry's growing need for quality security specialists. It's a sure bet that this exam is going to prove your subject matter knowledge and your ability to implement security processes and procedures.

REFERENCES

http://www.compTIA.org is the CompTIA Web site. All CompTIA certifications and exam information can be found here.

http://www.vue.com/comptia is the VUE registration site where you can register online for CompTIA certification exams.

http://www.2test.com is the Thompson/Prometric Web site where you can also register for CompTIA certification exams.

http://searchsecurity.techtarget.com provides a wealth of security-related information intended for security professionals in the enterprise.

CHAPTER 2

GENERAL SECURITY CONCEPTS

IN THIS CHAPTER

- Basic Terminology
- Access Control Systems Fundamentals
- Authentication
- Authorization
- Access Control Techniques
- Nonessential Services and Protocols
- Attack Methods
- Security Monitoring
- Auditing
- Test Tips
- Chapter Summary
- Review Questions
- References

This chapter begins our study of security concepts. It is an introduction into the broad spectrum of information technology in which security plays a role. In order for you to have a general understanding of how security systems and principles are implemented and function, it is important for you to understand a few basic security related terms. Consider it basic training for the security foot soldier. It might seem odd that this book begins with a sort of mini-glossary of terms. However, in order for you to have a better understanding of the material that will be covered, you need to have a few concepts behind you.

BASIC TERMINOLOGY

Please take a few moments to acquaint yourself with the following basic security related terms that will be used throughout this book. In the long run, you'll be glad you did and your exam scores might be the better for it.

- **Access:** To give the right or privilege to use an object or subject.
- **Access control:** Methods put into place that limit the access to system resources or physical locations.
- **Access Control List (ACL):** A list of user permissions and access rights, for example, Read, Write, and Execute, which are provided to the operating system. The OS uses the ACL to allow or disallow users or groups of users access to system resources, such as files or shares.
- **Access right:** Typically used to limit a subject's access to objects.
- **AES (Advanced Encryption Standard):** A 128-bit data encryption method adopted by the US government as a replacement for DES (Data Encryption Standard).
- **Attack:** The deliberate act of attempting to disable or paralyze a system or system resource.
- **Algorithm:** A set of instructions that typically make up a computer program or formula that is used to solve a problem or fix a reoccurring event.
- **Authentication:** A method used to verify the identity of a user or subject to a system. Authentication is typically a prerequisite for access to a system resource.
- **Authorization:** Permission that is granted to a subject to access or utilize a particular object, such as a file or folder.
- **Buffer overflow:** When an area of storage such as memory is overburdened and cannot function properly. This can lead to vulnerabilities in a system leaving it open to virus attacks, such as a backdoor virus.

- **Back door:** A weakness, penetrated area, or hole in an operating system's defenses. A back door in a system is opened so that an intruder or "back door" virus can enter.
- **Breach:** When the controls of a security system are penetrated and access to a system or resource is available to an outside influence or intruder, a breach has occurred. Typically known as a *security breach*.
- **Biometrics:** A computerized analysis of physical characteristics used to provide authentication or access; for example, fingerprint or retina scanning. Both are characteristic-based authentication methods.
- **Certificate:** A digital piece of information or a message that is used to verify that the requester of information is who they say they are. One who intends to send an encrypted piece of information applies for a certificate. The recipient of the encrypted message uses key to gain access to the information. A digital certificate is like an electronic credit card used to verify credentials for e-business transactions.
- **Certificate Authority:** The trusted provider or issuing organization that provides and guarantees digital certificates.
- **Cipher text:** Data or information that is encrypted.
- **Crack:** A program used to unscramble, decode, or decipher a password. Simply put, to break into a computer system.
- **Cracker:** One who breaks into a secured system with malicious intent. Crackers most commonly use brute force and dictionary attack methods as tools to figure out passwords.
- **Cryptography:** Method used to transform or encrypt plain text into an unreadable or unidentifiable format known as cipher text. In order for the encrypted text to be understandable, it must be decrypted.
- **Countermeasures:** Actions taken to reduce the threat or impact of a possible loss of data or property. Countermeasures such as passwords, antivirus programs, firewalls, or system policies can be implemented to reduce threat. They can also come in the form of physical security controls, such as padlocks or gates.
- **Decryption:** The process of taking information or data that has been encrypted and transforming or deciphering it into plain-text format. Cryptography methods are used for this conversion process.

- **Denial of Service attack:** An attack on a system or process that disallows or incapacitates the ability of a normal user or process to use the system or its resources. This is usually done by overtaxing the system with an amount of information, such as data packets, programs, or e-mail messages that the system cannot handle.
- **DES (Data Encryption Standard):** A widely used 56-bit symmetric private key encryption standard developed by IBM.
- **Digital signature:** An electronic version of a signature used to authenticate and identify the sender of information. Primarily used for identification purposes and the prevention of forgery.
- **Encryption:** A conversion process where plain text is converted to cipher text through cryptography algorithms in order to make information secret or unreadable.
- **Firewall:** Software, hardware, or a combination of both designed to prevent access to internal networks and resources from outside sources. A firewall is usually installed on a server that acts as a gateway or router. A firewall looks at data packets and screens them for validity.
- **Hacker:** An expert computer programming enthusiast who has the knowledge and capabilities to gain unauthorized access to secured computer systems and programs.
- **Hashing:** The changing or transforming of a set of characters into a shorter set or value of numbers. A hashing algorithm known as a *hash function* is used to disorganize values to make them more difficult to figure out. Hashing is often used with the encryption and decryption of digital signatures.
- **Intrusion detection:** Refers to a computer-related security management system that keeps track of authorized users and processes as well as identifies breach to networks and computer systems.
- **Nonrepudiation:** Proof that a transaction or contract occurred so that it is not denied at a later time. Digital signatures are a form or example of this proof.
- **OSI (Open Systems Interconnection) model:** A widely accepted seven layered reference model that identifies how data should flow from one location to another in a computer network.
- **Password:** A sequence of characters entered, typically without spaces, used to allow a user access to a system or resource. The entered password, which is a knowledge-based authentication

method, is typically compared to a security database on a host. If all requirements are met, the user is allowed access.

- **Plain text:** Data or information that is unencrypted.
- **Privilege:** An assigned user right. Privileges are assigned to users. Users can carry out tasks and certain systems functions based on their assigned privileges.
- **Private key:** The secret part of a key pair that is used to decrypt or unlock a previously encrypted or locked message.
- **Public key:** The not-so-secret part of a key pair that is used to encrypt or lock a message.
- **SATAN (Security Administration Tool for Analyzing Networks):** A program used to diagnose networks for security holes or weaknesses. SATAN is available as a freeware program.
- **Security policy:** A constantly updated set of rules and instructions that states how an organization will manage and protect itself.
- **Sniffer:** A network capturing and monitoring program used to troubleshoot network related issues, such as bottlenecks. A sniffer program analyzes network packets.
- **Smart card:** A card (typically plastic), which contains a processing chip and storage capabilities. A smart card is a token-based authentication device that allows its owner to gain access to a particular service, such as banking, parking, or gasoline services.
- **Spamming:** The proliferation or sending of unwanted junk mail.
- **Spoofing:** Falsifying one's identity in order to gain access. Pretending to be a valid user ID in order to penetrate a system protected by authentication methods.
- **SSL (Secure Sockets Layer):** A protocol that uses public and private keys to secure data transmitted over the Internet. A secured connection is established between a client and a server using SSL.
- **Token:** A small credit-card sized security device that contains an ever-changing identification code that allows its holder access to network resources.
- **Threat:** An action or behavior that could compromise security causing possible damage to resources.
- **Vulnerability:** A weakness in the design or structure of an operating system.
- **Virus:** A destructive program typically designed to spread to many computing systems and cause undesirable circumstances to occur.

Most computer viruses come in the form of e-mail attachments or are disguised in programs that are downloaded from the Internet.

- **Worm:** A computer program (virus) designed to duplicate itself again and again until it exhausts system resources.

Two of the best sites on the Internet for referencing security-related terms and their explanations are *http://www.sans.org/newlook/resources/glossary.htm/* and *http://www.securitypanel.org/glossary.html/*.

Now that you have a basic understanding of a few of the important security-related terms that will be covered and referenced throughout this book, we will begin our study of security basics with a conceptual overview of access control systems fundamentals. Finally, we will explore the most commonly used methods and techniques implemented to attack a network or computer system.

ACCESS CONTROL SYSTEMS FUNDAMENTALS

Access control defines the set of procedures used to restrict and monitor access to various types of sensitive information or resources. Access control can be implemented by hardware, software, or by IT administrators to do the following:

- Identify users
- Track users' use of resources
- Allow or deny access to those resources

Access is granted, limited, or denied, based on one's identity or membership in a predefined group through which privileges are inherited. The time of day or location of the user can also play a role in their level of access. The Security+ certification exam will require your knowledge of the different techniques and methodologies used to implement access control in an enterprise environment.

The most common type of access control in effect is witnessed when a user is denied access to a password-protected network share. Either the user wasn't specifically granted access to the resource or another restrictive measure is being enforced. Next, we will explore some of the important access control models from which many access control methods and techniques are based.

ACCESS CONTROL MODELS

If you have studied networking fundamentals, you are most likely familiar with the *OSI (Open Systems Interconnection)* model. The OSI model is used as a theoretical reference for programmers and developers to use as common ground for developing and implementing new programs and network related protocols and devices. (The OSI model will be described in detail in Chapter 4). Just as the OSI model is used as a theoretical approach to networking, access control models are used as theoretical approaches to the various access control methods we will discuss. Although you should focus your attention on the specific access control techniques detailed later in this chapter, it is important for you to familiarize yourself with the basic security models from which access control techniques are derived.

 It is likely that the exam will ask you to identify the type of access control used for a particular security model. For example, The Bell-LaPadula model is based on discretionary access control.

The most common access control models are as follows:

- **Bell-LaPadula (B-L model):** A mandatory access control model developed to control and protect government and military information and data. The Bell-LaPadula model was the first mathematical security model used to address security, modes of access, and a set of rules for assigning security access rights. This access control method is a hierarchical structure where access is based on the assigned rights and classifications of subjects and objects. With this approach, subjects and objects are assigned different levels of security. A subject can access only objects based on a subject's security clearance or level. This model also supports the ability to verify access rights by checking an active matrix. This form of access control is known as *discretionary access*, which restricts a user's access to an object. For example, a file or folder.

- **Biba:** In 1977, the Biba security model was created to address some of the particular weaknesses in the Bell-LaPadula model. Specifically, the Biba model addresses the problem concerning the ability of a subject or user with a lower security level rating to write to a subject's information with a higher security level or clearance. In simple terms, if User B has a lower security level than User A, User B should not be able to write over User A's information.

- **Clark-Wilson:** The Clark-Wilson security model was developed in 1987. Its main focus is to protect the integrity of data through the use of secured programs. This model is concerned primarily with the internal and external consistency of data.
- **Non-interference:** A mathematical technique developed for high-level security systems. Non-interference was designed as a tool for analyzing or testing the security of a computing system.

If you are interested in learning more about security models, a wealth of information is available for free on the Internet. Simply go to your Web browser and search *Security Models*. It is unlikely that the current Security+ exam will drill you on the fine details of the mathematical equations that make up these models. However, it is likely you will have to know what type of access control they use.

AUTHENTICATION

User authentication is the first step in accessing controlled system resources. Once the system knows who you are, it can provide you with the resources that you've been granted access to. This element of access control enables administrators to trace user activities and hold users responsible for their actions. A user name or identification, in conjunction with a password, is the most widely used method of authentication. User IDs must be unique on any given system for them to be effective. A user ID is not meant to be descriptive of a particular job function; it simply asserts one's identity and sets the stage for a system of access, accountability, and monitoring.

An individual can be identified on a system by entering a user ID and password, swiping a magnetic card, keying a code on a keypad, or by some physical trait unique to that user. *Biometric authentication* determines one's identity by a fingerprint or handprint, retina scan, facial scan, or voice recognition. Combining a physical trait or a possession (magnetic card, key, and so on) with a password or PIN (personal identification number) can further enhance security. This type of two-factor authentication is also known as *Strong User Authentication (SUA)*. Remember, authentication itself does not determine the specific rights of a user; it's just the method of ensuring that the user is who they claim to be. Next, we will explore different types of authentication.

PASSWORDS

Being the weakest link in any security system, it's important to know what makes passwords stand up well against *brute force* (or password guessing) attacks. Because many password-cracking programs use multilingual dictionaries and common name lists to get their ideas, words or names found in such references should never be used. The name of your company or organization should never be used. Too many repeating characters within a password should be avoided as well. A lengthy, nonsensical combination of mixed-case letters, numbers, and symbols (if allowed by the system) is your best bet. This makes a password easier to forget but there are a couple of tricks to help you remember. Examine these passwords: GreaterThan> therest; $tr0ngPa$S, Apri1inP@ris, and ParenTHETICALone(1). You get the idea. Make passwords as long as the system allows. Your password should also be changed on a regular basis. Most systems can be programmed to make users change their passwords at predetermined intervals but this should be done, regardless. The more complicated your password, the less often you will need to alter it. Most systems also keep track of unsuccessful log-on attempts and lock out user accounts after a certain number of failed tries. Don't ever write down passwords anywhere, especially under your keyboard!

In Law #5 of their "Ten Immutable Laws of Security," Microsoft™ states, "Weak passwords trump strong security." (This Microsoft quote can be found online at: *http://www.microsoft.com/technet/treeview/default .asp?url=/technet/columns/security/essays/10imlaws.asp.*) In other words, no matter how sophisticated your server, a malicious user need only obtain your password to cause damage, steal sensitive data, or play "you;" and you are exactly who will be held responsible.

User IDs are out there for the entire world to see and passwords can be guessed, cracked by software, or stolen. Although simple to administer, the future of enterprise security lies within the systems and devices that go beyond basic password authentication.

The following methods and criteria should be used to create strong passwords:

- A strong password should be a minimum of seven characters in length.

- The password should contain a combination of upper- and lowercase letters.
- There should be at least one number contained in the password.
- The password should contain at least one of the following characters: !@#$%^&*.
- A password should never be a word from a dictionary, a person's name, family name, phone number, birth date, or favorite phrase.

Here's a final note on passwords: All of the money that is spent by a company for security personnel, security mechanisms, and software can prove basically worthless if network users and company employees can't remember their network passwords and post them on computer monitors, under keyboards, or in desk drawers. Get the message? What good are all of the high-tech security mechanisms if basic security policies are not followed? We will discuss policies in Chapter 6.

TOKEN-BASED AUTHENTICATION

Token-based authentication is among the most widely used SUA systems. This method achieves a high level of security because the access information is carried in a portable unit. A perfect example of this technique in use is the ATM card. When you go to withdraw cash, you insert your card, which is always safe in your possession. You then enter a PIN, which stays safely tucked away in your mind. Automatic banking would have never gained the popularity it has if it weren't for this type of two-factor routine we go through to prove we are who we say we are. Can you imagine the nightmares it would create if all you needed were someone's name and PIN to empty their bank account? The ATM scenario is an example of a *memory token*, that is, the card holds data but does not process it. Many organizations are finding themselves in a situation where their data is as valuable, if not more valuable, than cash.

Another type of token-based system, such as the RSA SecurID®, uses a *smart token*-enabled, battery-operated device. Like the ATM, a PIN is entered into the system each time access is requested. The smart token system, however, contains access information **and** has the benefit of processing capabilities. These devices can contain internal clocks synchronized with their parent servers, adding another level to the security. The server compares its clock to the device clock, and if it gets a match, you get access. In addition to this time-synchronous routine, these devices share a uniformly updated seed (or random number) with the server introducing yet another step in the algorithm used for authentication. Again, if this random number matches, you're in. All of these steps equal a nearly fool-proof method of identifying an individual. One of the advantages of this type of authentication is its potential portability. Smart token systems can be integrated into devices already in your pocket, such as a handheld device or mobile phone.

KERBEROS V5 AUTHENTICATION

Developed by the Massachusetts Institute of Technology (MIT), *Kerberos* is a network authentication protocol that employs strong cryptography enabling secure client/server communications over an insecure connection. Known as a *distributed authentication service*, Kerberos allows users to identify themselves on a network without exchanging data (passwords) that a third party could intercept. This type of cryptographic authentication shields the identification process from prying eyes. It's important to understand that the Kerberos protocol does not protect all data exchanged between two systems. It encrypts only messages sent between "Kerberized" applications or software that was built or modified to take advantage of this protocol. Think of Kerberos as simply another type of language used in the authentication process. It doesn't provide the authentication itself; rather, it provides a mechanism through which authentication information can be exchanged securely.

The Kerberos authentication protocol is outlined in detail in the Requests For Comments (RFC) 1510. It uses a string of encoded messages and the issuance of special tickets to verify the identification of the user in question and allow or disallow a user's access to objects. Appropriately named after the three-headed dog that guards the gates of Hades in ancient Greek mythology, Kerberos works as follows.

The first step in this process is, as always, the request for access. Using a password or smart card, the user attempts to access a service. The rest of what follows is transparent to the user. Upon receiving the request, the *authentication server (AS)* issues a *ticket-granting ticket (TGT)* to the client. This encrypted ticket includes the user's password and a random seed representing the requested network service. The client machine then returns the ticket to the *ticket-granting server (TGS)*, which may or may not be the same machine as the AS. The TGS then issues a service ticket to the client. Once the client machine possesses this service ticket, the ticket can be used to request a service. The service ticket verifies the user's identity to the service. It is the credentials that the service uses to allow or deny access by a client machine. Furthermore, the TGS time-stamps service tickets enabling further requests to be made by the client without repeating the process. The expiration time is typically within five to10 hours of issuance. However, if the network is on a large scale and the user attempts to access resources maintained by a different AS, reauthorization will be necessary unless the system supports Single Sign-On.

 For the exam, be aware that a Kerberos server is used for authentication and encryption.

More information regarding Kerberos and other Microsoft security related technical information can be found at the Microsoft MSDN Web site at the following URL: *http://msdn.microsoft.com/library/default.asp?url= /library/en-us/security/Security/using_security_packages.asp/.*

SINGLE SIGN-ON

Many networks today have multiple servers in different buildings, or indeed different cities, running several operating systems and providing users with tons of resources. The more passwords a user needs to remember in order to access all that data, the higher the likelihood of those passwords being recorded in places other than their heads; for instance, places such as beneath their keyboards (tsk, tsk). Also, users will select the same password for access on different systems. If just one of those systems has an insecure authentication method, an attacker can intercept the transmission, detect the password, and gain access to other "secure" systems.

SSO provides a secure way for users to be authenticated just once while enabling enterprise-wide access to data. It also simplifies the administration of tasks such as disabling a user account. In an SSO environment, the network administrator, whose time is surely tight, can disable just one account and be assured that there's no lingering access on a machine somewhere in the network. There are many methods available to enable SSO in one form or another and many are costly and difficult to implement. They can benefit us by creating a more efficient authentication process but there are precautions to take to prevent hackers from exposing SSO weaknesses and killing many resources with one stone.

As we've discussed, strong passwords are a must. This is even more important to consider when one-stop authentication is in effect. Systems implementing SSO should enforce the selection of stronger passwords. They should demand alphanumeric, mixed-case passwords of at least seven characters. An SSO environment should also employ two-factor authentication because a compromise of security could be more damaging.

Exactly how SSO is implemented depends on whose product you use to deploy it. Novell®, Inc. and Passlogix®, Inc., for example, have combined their technologies to create what they call the Novell Single Sign-on (NSSO) 2.1 bundle. This implementation introduces a low overhead method of adding SSO to an existing Novell network. Novell claims that their process ensures secure SSO capabilities providing access to UNIX hosts, Web sites, desktop applications, and in-house applications while protecting passwords in a patented repository. Another system taking a public/private key approach to SSO is called *SESAME (Secure European System for Applications in a Multi-vendor Environment)*. SESAME's development began with the open systems standards work of the *ECMA (European Computer Manufacturers Association)*. This implementation is not a complete product but rather a foundation upon which vendors can develop other SSO technologies. Concepts within the SESAME project have been used to develop an expanded version of Kerberos that supports some of the ECMA's security architecture.

BIOMETRICS

Character-based authentication methods allow or disallow access to systems, resources, or physical locations based on physical characteristics. *Biometrics* is a combination of science and technology that is used to gather and measure human characteristic information from a subject and use that information as a means to allow or disallow access. Some of the most popular characteristics that can be measured and used with scanning devices through the use of biometric technology are handwriting, hand imprints, fingerprints, and retina, iris, and voice patterns.

A retina scanner is currently the most secure biometric device available. Fingerprint devices and signature scanner devices are the most widely accepted forms of biometric implementation. However, they are not as secure as retina scanning devices.

Biometric security is considered to be among the most secure methods to allow or disallow access. The high demand for this secure technology can be seen in just about every aspect of our daily lives. Bank and purchasing transactions can utilize this technology. Government, military, and corporate environments will employ this technology wherever the need for a reliable personal identification system is required. However, the use of biometrics is still considered to be in its infant stage of development.

Biometric authentication can currently be implemented into a network infrastructure to allow secure access to applications, domains, workstations, and servers. It can be used locally or as an authentication method for remote access. Biometric solutions can also be implemented with other access methods of authentication such as smart cards, encrypted keys, and digital signatures.

Most believe the future of network security lies within the biometric realm. Many police organizations already realize the benefit of using retina scanning as a means of identification.

AUTHORIZATION

Authorization is the second step in accessing protected data or objects. After a user has been identified, another set of standards is in place to determine which resources should be made available to the user, and what exactly the user may do with those resources. Most of the systems that we've just covered are based on this two-step process. Each has its own way of authenticating users, and in turn, each has a unique method of checking the access privileges of the resource being accessed. In addition, each system has a unique way of delivering these privileges in a secure manner. For instance, the types of file system access rights granted in the authorization process include but are not limited to the following:

- **Read:** Allows reading of files or listing the contents of directories.
- **Write:** Allows writing to files or adding files to directories.
- **Execute:** Allows the execution of program files.
- **Append:** Allows the addition of data to files or placement of subdirectories into directories.
- **Delete:** Allows the deletion of files or directories.

A user can also be granted any combination of these and other rights, which different operating systems label in different manners. As you'll see in the next section, system administrators, organizational policies, or even other users can determine these rights.

ACCESS CONTROL TECHNIQUES

Now that we've examined the methods of secure authentication and authorization, deeper discussion is required to explain the many system-wide policies (or access control techniques) that dictate the way these parts pull together as a team. Planning an effective security solution involves the selection of a *security policy*. The security policy you choose will characterize the behavior of your information security system regarding control over access and the distribution of resources. There are different access control techniques that achieve this goal in varying ways. Some restrict access based on users' identities while others focus on what role the user plays in an organization. In this section, we'll discuss the intricacies of these different techniques and clarify their objectives.

 It is very likely that the exam will drill you with questions on these access
control techniques. Be sure you know them very well.

DISCRETIONARY ACCESS CONTROL (DAC)

Discretionary Access Control restricts access to resources based on the identity of users and/or groups of which they are members. This technique is referred to as *discretionary* because it also allows users to pass on access permissions to other users within the organization. For instance, a set of users could have permission to grant or deny access to a file they own, such as a text document. They can select the users that need access to it based on their own opinion of who should have such access; based on their own *discretion*, in other words. Furthermore, they can decide who should have only read access and who should be able to read **and** write to the file in question. One drawback of DAC is that its effectiveness is limited by the ability of users to make concrete security decisions. It's very possible that an inexperienced or careless user could inadvertently grant full access to files or objects to an entire organization's users. Although also employed by other operating systems, to gain a better understanding of how this works, we'll examine Microsoft's implementation of DAC and the parts that provide its functionality.

MICROSOFT WINDOWS NT/2000 DAC

At the heart of Windows NT/2000 DAC is the Access Control List (ACL). ACLs come in two types: *discretionary* and *system*. Simply stated, an *ACL* is a list of who may access an object and what rights to that object they possess. The *Discretionary Access Control List (DACL)* is maintained by the owner of an object and determines the specific users or groups who can access the object. An ACL is kept current by the addition of Access Control Entries (ACEs). An *ACE* contains a *Security Identifier (SID)* and the set of access rights that identifies the trusted user for whom the rights are denied, allowed, or monitored. The *System Access Control List (SACL)* creates the audit messages when protected objects receive access requests. The network administrator typically controls management of the SACL. The Windows NT/2000 DACL also includes a group of *Discretionary Access Control Entries (DACEs)*, which consist of, among other data, an object's access mask. The access mask indicates the kind of access requested, such as Read, Write, or Execute.

ACLs are usually large, complex databases that are subject to the use and abuse of a wide range of users. There is also high overhead involved in maintaining and updating a DAC-based security policy. For these reasons, it can be argued that DAC is more difficult to implement and less secure than other techniques of controlling access. With the exception of added user-controlled access granting capabilities, *Mandatory Access Control (MAC)* achieves the same or higher level of security while being less difficult to implement and administer.

MANDATORY ACCESS CONTROL (MAC)

Mandatory Access Control, also called *nondiscretionary*, uses a centralized approach to restrict access to data based on the sensitivity of the data in question. There have been several variations of this access control technique published since its inception, but it was Bell and LaPadula, as explained earlier in this chapter, who in the 1970s originated the concept of MAC. This technique assigns hierarchical, multilevel sensitivity *labels* to users and data (as in the military—unclassified, confidential, secret, and top secret). In this labeling system, user labels are referred to as *security clearances* while object labels are referred to as *security classifications*. The idea is that if you've been labeled with top-secret clearance, for instance, you'll be able to view documents classified as top secret.

Opposed to DAC, MAC puts into the hands of system administrators the decisions regarding who sees what. This technique is often used in situations that require exceptionally high security standards, such as governmental agencies. MAC is also highly compatible with an organizational setting because its policies mimic those of the preelectronic methods of document security. As with other security techniques, MAC can be based on a wide variety of traits and responsibilities that the users under its enforcement possess. Next, we will discuss a few important specialized implementations of MAC-inspired access control techniques.

ROLE-BASED ACCESS CONTROL (RBAC)

RBAC is essentially a type of MAC, but rather than sensitivity traits, its access control evaluations are based on the role users play within an organization. RBAC has recently been enjoying growing attention as an alternative to DAC-based systems because of its reduced administration and lower level of complexity. Applications employing RBAC provide a mechanism of formulating the system's structure to compliment the existing structure of an organization. RBAC enables organizational planners to put an access control policy into place that the system must abide by, rather than a policy that changes with the whim of a user or administrator. In other words, within an RBAC system, the organization **owns** the resources that are protected. This introduces a higher level of security not possible with other techniques. In the simplest of terms, RBAC can be used to enforce easily the minimum amount of privileges a user needs.

Don't confuse role-based access control with *rule*-based. As in DAC, rule-based access, which also uses ACLs, simply weighs user rights against object-specific security rules to grant a user access.

LATTICE-BASED ACCESS CONTROL (LBAC)

LBAC uses a mathematical formula, or *lattice*, to determine whether or not access to a resource should be granted. LBAC was developed to manage the flow of information from one security label to the next (or one object to the next). The idea of LBAC is essentially to combine discretionary **and** mandatory access control system fundamentals in a way that compels this flow of information. The formula, which is only summarized here, works as follows.

When a secured object receives an access request in a LBAC system, the first step is to check what's called the *discretionary access matrix*, which can be altered by users. Then, the system-controlled MAC guidelines are referenced. The lattice is then referenced and its formula is applied to the clearances and classifications of the respective users and objects. Its result is what's used to grant or deny access. This two-step process provides the benefit of high-level security while enabling a degree of user discretion with regard to access inheritance.

TASK-BASED ACCESS CONTROL (TBAC)

TBAC is more than just a bathing suit. It's actually an exciting, relatively new technique for implementing what's known as *active* security. TBAC bases its access decisions on the current state of works in progress. Suitable for distributed computing environments and working at the application level, the TBAC technique applies a continually updated access control mechanism to work flows or *tasks* as they near completion. In other words, a TBAC system could be configured to disallow access to documents within a specific area of a project's resources at such a time when it's determined that the specified area has reached a state of conclusion.

As previously stated, MAC-based techniques of authentication use a centralized approach to access control. That is, the user names, access rights, and permissions are all stored on one central server. This is in contrast to the Windows NT decentralized method, which uses a system of trusts and domains. In such a system, access information is stored on the many computers that comprise the network and each system plays a part in determining who has access to what.

There are two more notable implementations of centralized access control that warrant our discussion: RADIUS and TACACS.

REMOTE AUTHENTICATION DIAL-IN USER SERVICE (RADIUS)

RADIUS is a scalable, client/server-based UDP protocol used to transfer authentication and authorization data between a dial-in client and a server. RADIUS enlists the use of two servers: one for authentication (*RADIUS authentication server*), and one that acts as a *centralized* database for user profiles (*RADIUS accounting server*). The RADIUS authentication server, which can be a firewall, router, or dial-in server, requests accounting information from the RADIUS accounting server when a user attempts to log on. The accounting server then answers with an encrypted response telling the requesting server what access to provide. The RADIUS protocol supports PAP, CHAP, PPP, and UNIX login authentication methods.

TERMINAL ACCESS CONTROLLER ACCESS CONTROL SYSTEM (TACACS)

TACACS is a dated, remote authentication protocol typically implemented on UNIX servers. It permits a remote access server to query an authentication server to establish users' access privileges. TACACS works in a similar manner to RADIUS in that it contacts an authentication server over the network and requests verification information from a centralized user database. Unfortunately, TACACS does not encrypt its messages to and from the server. Cisco© has developed a newer, extended version of TACACS called *TACACS+*. This extended version adds new features, such as encryption and extended logging functionality.

INFORMATION SECURITY CONTROL TYPES

In conclusion, let's talk about some of the categories of security control from an administrative point of view. There are essentially three categories of information security controls:

- Physical
- Logical
- Administrative

Physical controls entail the use of instruments such as locks, fences, motion detectors, battery backups, and burglar/fire alarms. They are the material systems and devices that protect assets from theft, fire, or other means of destruction.

Logical controls refer to the systems such as access control software, antivirus software, communications hardware (routers), passwords, and smart cards. They represent the systems in place that prevent unauthorized access to digital information.

Finally, *administrative* controls are the personnel-related mechanisms for managing people's behavior. These include security training, background investigations, mandatory vacations, and performance assessments. These categories are also broken down into the five following subcategories that more specifically define their intent:

- **Preventative:** Avoid violations before they occur.
- **Deterrent:** Discourage violations from occurring.
- **Detective:** Investigate violations that have occurred.
- **Corrective:** Remedy violations that have occurred.
- **Recovery:** Restore lost resources from a violation that has occurred.

NONESSENTIAL SERVICES AND PROTOCOLS

Many of the operating systems available today can be made to be somewhat secure. Unfortunately, it is common for many default installations of operating system software packages to leave systems vulnerable to outside attack. For security purposes, it is essential that you stop (disable) any un-needed services and remove or *unbind* any unnecessary protocols that are not needed by a system that has direct connection to an outside influence such as the Internet.

If you are using the Windows operating system and have an external connection to the Internet, you should consider removing the following protocols from your external NIC cards TCP/IP protocol bindings. This can be accomplished through the Windows Control Panel:

- Server
- Workstation
- NetBIOS interface

You should also consider removing or disabling the following services from a system if they are not needed. Keep in mind that disabling particular services might render a system useless for particular tasks:

- Computer browser service
- IIS Admin service
- FTP Server service
- Spooler service
- Netlogon service
- DHCP (Dynamic Host Configuration Protocol)

It is likely that the exam will ask you what unused services should be removed or disabled from a system to reduce the risk of malicious attack.

From a network as well as operating systems perspective, the following items should always be taken into consideration when hardening an environment from outside destructive forces:

- Always apply the most recent hot-fixes and service packs available from your operating systems manufacturer.
- Block all TCP/IP and UDP ports that are not needed for network traffic. As a general rule, block TCP port 139 and UDP ports 137 and 138.

 It is likely that the exam will target this issue. The blocking of these ports should be on the first page of any Security 101 book or procedure.

NOTE

- Enable a strong password structure.
- Enable auditing and logging .

ATTACK METHODS

You hear about them on the radio, you read about them in the newspapers and in newsgroups. They are receiving more and more media attention than ever. If you are unlucky enough, you might experience them first hand on your home PC or your office workstation. They are computer attacks! Computer attacks, intrusions, and information theft are well on the rise. If companies and PC owners alike do not invest in the proper resources such as training, information security personnel, and proper hardware and software to prevent the continuing growth of this illegal activity, there might soon be nothing left but metal boxes and electronic circuitry. Your company and your country need you to get educated and Security+ certified!

This section concentrates on the most common types of computer attack methods that you are most likely to encounter on the current Security+ exam. In order for you to protect your network or PC, it is first important to gain an understanding of the methods intruders use to break into your system and ultimately, your privacy. We will concern ourselves with intrusion detection and protection later in this book. For now, here are the most common attack methods used to bypass security access controls and compromise your information or system.

 It is very important that you understand each of these attack methods. It is likely that the exam will present you with a scenario. You will then have to choose which type of attack method is being implemented.

BRUTE FORCE ATTACK

A *brute force attack* comes in the form of a program that uses trial and error methods to guess or figure out passwords, encrypted keys, or PINs. This form of attack uses all possible combinations of characters and words in order to figure out a solution. Many administrators use software called LOphtcrack from Lopht Heavy Industries, which implements brute force algorithms to crack or test their own passwords. Brute force is considered to be a reliable but very exhaustive and time-consuming attack method. An excellent way to avoid this type of attack is to implement an *OTP (One Time Password)* authentication method. With OTP, a password can be used only once to access a resource. After that, the password is no good.

BIRTHDAY ATTACK

A *birthday attack* is the name used to identify a form or class of brute force attack that attempts to resolve a certain class of cryptic hash functions. The birthday attack gets its name from the *birthday paradox*, which states that for every 23 people, the probability that two will share the same birthday is greater than half. The birthday attack uses a formula based on this concept to calculate often collisions in hash functions.

DENIAL OF SERVICE (DOS) ATTACKS

A *Denial of Service (DoS)* attack is most commonly launched as a malicious attack with the intent of disabling or removing computer services or resources that are normally available. Although sometimes unintentional or accidental, most DoS attacks are intended to stop or destroy network related services such as Web sites or e-mail. Typically, this type of attack is designed to render a network or related service useless by flooding the network with worthless or useless network traffic. The following six attack methods are types of DoS attacks:

- **Buffer overflow attack:** A *buffer* is a temporary data storage area, usually *RAM (Random Access Memory)* that holds a predefined amount of data that is shared by programs and devices. If a buffer gets full, data that is meant to be held in a particular buffer might overflow or go elsewhere, possibly overwriting data held in other buffer storage locations. Knowing this, attackers can target specific systems or network nodes with malicious packets they suspect might have been developed with weaknesses. A buffer overflow is the most typical type of DoS attack and can render a system or its resources useless.

- **SYN attack:** When a session or request from a *TCP (Transmission Control Program)* sender or client and a network file server or host is started, a *SYN (synchronize/start)* request is sent to the host from the client. The host must answer with a *SYN acknowledgement (ACK)* before a complete connection is established. This request and acknowledgement between client and host is known as *handshaking*. There is a small buffer that exists on the host that handles the constant handshaking or transfer of acknowledgement packets before sessions are established completely. This buffer contains a *SYN field*, which identifies the sequence in the message exchange. Possible attackers can send packets to the host systems requesting a connection and fail to respond to the SYN acknowledgement. This type of attack leaves the SYN (synchronize/start) packet request in the buffer blocking other legitimate requests from being acknowledged. In other words, the attacker stops or impedes the ability of the server to establish or hand out sessions to clients. To best prevent this type of attack, administrators should implement the following operating system changes:

- Increase the size of the connection buffer (SYN ACK queue).
- Decrease the time-out waiting for the three-way handshake.
- Obtain and employ vendor software patches that detect and avoid this type of attack.

- **Teardrop attack:** *IP (Internet protocol)* uses a special technique to divide large network packets into what are called *fragments*. When fragment reassembly code does not properly handle the repackaging of these fragments into packets, a weakness is exposed. An attacker can insert code into these fragments before they are properly put back together. This method can exploit networks and cause systems to fail. Special vendor patches are the recommended solution to this weakness.

- **Ping of death attack:** The IP protocol places a sending packet size limitation of 65,536 bytes per packet sent on an IP network. As mentioned earlier, IP can break down a packet into IP fragments or segments before they are sent out on a network. A known weakness of this method is that fragmented packets can be reassembled to equal a packet size greater than the 65,536 IP limit. Operating systems that attempt to receive these oversized packets can be overburdened with the reassembly of these confusing packets and crash. In order to hide the identity of the attacker, spoofing (mentioned again, shortly) can also be implemented during the type of DoS attack. Once again, operating system patches and protecting a network from *Internet control message protocol (ICMP)* broadcast messages at a network router or firewall are the recommended ways to protect a network from this attack method.

- **Smurf attack:** This is another DoS attack where an attacker or perpetrator floods an IP network with echo replies in response to ping messages sent out by a victim. The packets that flood the network are sent to all possible IP nodes on a given network congesting the network until it is useless for normal IP traffic. The attacker typically uses a technique known as *spoofing* to carry out this attack. The packets are spoofed, meaning they are sent out with someone else's return address. The best-known method of defending against a Smurf attack is to disable IP multicast broadcast addressing at each router on each subnet (sub-network) within a corporate network.

- **Land attack:** This is another DoS attack where a malicious attacker attempts to set up a TCP session with a server computer. If the server establishes a session with the attacker, the server will fall into a closed loop and will have to be rebooted. This is also a form of IP spoofing.

DICTIONARY ATTACK

This type of attack typically uses a predefined list of words such as those found in an English dictionary. The attacking program runs through the list of words until it finds a match to the password it is searching for. Today, a brute force attack is used more commonly to calculate a password or figure out combinations of words in place of this outdated method of using predefined lists. A successful dictionary attack can be avoided by keeping password databases and lists in secured and secret areas of a network. Also, passwords stored in a database should be stored in a one-way hashed fashion or form. A *one-way hash* is a mathematical function that is difficult to reverse; it takes a variable-length input string and converts it into a fixed-length binary sequence. In simple terms, it makes the passwords hard to crack.

MAN-IN-THE-MIDDLE ATTACK

With this type of attack, the attacker uses a program that places them between the sender of information and the receiver, appearing to the sender as a client, and appearing to the receiver as a server. While "in the middle," the attacker can intercept data and information and replace it with bad or destructive information.

 A Man-in-the-Middle attack is not considered a DoS attack. You might be expected to know this on the exam. Be able to differentiate the individual attack methods as well as the attack types.

NOTE

REPLAY ATTACK

A *replay attack* is considered a form of *degradation of service attack* as opposed to a Denial of Service attack. With a replay attack, small bursts of traffic are sent from multiple locations to a single host. This inundates the host with traffic. The functionality of the host declines over time until it is basically unable to provide resources and services.

In order to combat and prevent this type of attack, the IETF (Internet Engineering Task Force) Internet Protocol Security IPSec standard uses an anti-replay protocol. This protocol makes it virtually impossible for packets to be intercepted by code inserting hacking programs that hijack packets transmitted between source and destination systems. In other words, use the IPSec standard to provide packet-level security and avoid this type of attack.

TCP/IP HIJACKING

In networking security terms, *hijacking* means taking control of a communication session while it exists. There are several types of hijacking techniques used by modern-day hackers and would-be thugs. A man-in the-middle attack (mentioned earlier in this chapter) is a type of hijacking attack. There are also DNS (Domain Name Server) hijacking techniques.

With a basic TCP/IP hijack, a perpetrator can gain control of a communication session if his or her attacking program can acquire a *random initial sequence number (ISN)* that is used by two systems to start a packet transmission sequence. This is an older form of hijacking that has supposedly been addressed in IPv6 (Internet protocol Version 6).

For an excellent description of TCP port hijacking, please see the information located at *http://cs.baylor.edu/~donahoo/NIUNet/hijack.html*.

PASSWORD CRACKERS

Password crackers are programs such as Lophtcrack that can figure out easily passwords that are secret or encrypted. Password cracker programs can employ (for free) attack methods such as brute force, dictionary attacks, and other methods to break passwords.

SNIFFERS

Sniffers are programs or devices that gather network packets. They can be used for legitimate network administration purposes, such as monitoring network traffic, or for destructive and illegal purposes, such as extracting passwords, user IDs, PINs, credit card numbers, and other personal information. Sniffers are very dangerous tools in the wrong hands because they can be placed or inserted just about anywhere in a network and can go un-detected easily avoiding intrusion detection implementations.

SPAMMING

Spam is the unsolicited proliferation of electronic advertisements, junk e-mail messages, and newsgroup postings on the Internet. "You've got spam" has become a familiar phrase among e-mail junkies. In short, it is considered rude and indecent to send or proliferate spam. Spam is often compared to those obnoxious phone calls you receive, while enjoying your dinner, asking you to change your phone service or purchase another wonderful plastic "I want to go further into debt" credit card.

SPOOFING ATTACK

Spoofing is a technique used to pretend to be someone or something that one is not in order to gain access to a resource that belongs to another. Spoofing is most commonly used to fake an e-mail or IP address. As mentioned earlier in this chapter, most DoS attacks involve some sort of spoofing technique.

SOCIAL ENGINEERING

Social engineering sounds like a pretty constructive and harmless set of words, doesn't it? Don't be fooled! Social engineering is like a spoofing attack. It means to fool someone or something into allowing unauthorized access with intent to cause destruction or obtain information illegally. Social engineering is based on the theory that humans are generally trustworthy. With this in mind, the social engineer or hacker typically uses a computer or phone to pursue their victims into providing information.

Another form of social engineering is known as *reverse social engineering* whereby a hacker pretends to be an authoritative figure. While acting as the superior authority, the hacker is asked questions by company employees. Using this technique, the hacker can then gain useful information from the employee to access records, information, or bypass security measures. The receiving of fraudulent e-mail is a perfect example of social engineering.

 You must know the difference between spoofing, spamming, and social engineering. It is likely that the exam will bombard you with questions in an attempt to trick you here.

NOTE

SECURITY MONITORING

Network and computer monitoring is essential to the health of a corporate or home computer network. It is imperative that monitoring be implemented in order to keep the eavesdroppers, hackers, and internal malicious network users at bay. *Monitoring* is the process by which a system or network is evaluated over an extended period of time using *Intrusion Detection Systems (IDS)* (covered in Chapter 4) that can be either hardware devices or software. *Vulnerability scanners* can be used to assist you in detecting which areas of your network are weakest and need a watchful eye.

Today, it is fairly easy and inexpensive to implement monitoring through the use of network packet sniffers. There are many monitoring tools available to network administrators and network security personnel from many different vendors. Some of the most useful monitoring tools and information regarding security systems can be found at Cisco.com, specifically, *http://www.cisco.com/warp/public/44/solutions/network/security.shtml/*.

AUDITING

Auditing is a way of tracking predefined events of a user or group of users on a particular computer system. Most modern day operating systems offer a built-in auditing tool or utility for tracking certain events. Windows 2000, for example, offers the ability for administrators to audit, log, and track the following events:

- Account log-on
- Account management
- Directory service access
- Log-on events
- Object access
- Policy change
- Privilege use
- Process tracking
- System events

Auditing too many and unnecessary events on a workstation or server computer can hamper drastically the system's performance. In order to minimize the threat or risk of threat on a system, Microsoft offers several recommendations for certain important events that should be audited.

Enabling failure of user log-ons and log-offs can assist with identifying the threat of possible random password attacking programs or hacks. To avoid the misuse of user privileges, you should enable auditing for security change policies and certain system events, such as system shutdown and restart. After auditing has been enabled in Windows NT or Windows 2000, the results of the audited events can be viewed in the Windows Event Viewer.

As mentioned earlier in this chapter, different operating systems offer various ways to enable, track, and view the results of auditing. If you are responsible for protecting the welfare of your networked environment, you should have a good working knowledge of the auditing tools available within the operating systems you are running.

Auditing tools such as those used in Windows 2000 and UNIX can track system and user events as well as abnormalities. These tools are considered a detective control. These tools produce log files that should be reviewed continuously. In UNIX, a network administrator should audit and log the following operating system conditions and events:

- The UNIX kernel
- Lib files
- Bin files
- Use of Setgid
- Use of Setuid
- Changes made to the */etc/password* and *.rhosts. files*

As a general rule, audit logs should be kept for at least a year. This retention period can vary depending on your company's needs and policies. You should also remember for the exam that log files should be encrypted if they are ever transmitted over a network. They should always be kept secure to avoid deletion, modification, or destruction.

TEST TIPS

As stated in Chapter 1, a "Test Tips" section is located at the end of each chapter. These Test Tips are a brief review of some of the important topics covered in the particular chapter that might be targeted by the Security+ exam. Plus, as an added bonus, there might be a few tips that you have not run across that could be helpful in your preparation. If you have trouble with a topic, it is suggested that you go back in the chapter and review it again until you understand it completely.

Also remember that there is a wealth of information available on the Internet regarding most of the topics described in this book. If you are just interested in knowing more about a subject, use every available resource. This book is designed as a preparation study resource. It is not designed to teach you everything there is to know about security. You will need a library for that.

- √ An operating system can use an *access control list (ACL)* as an authentication method to see what rights a user has to a certain object such as a file, folder or network share.

- √ With *discretionary access control (DAC)*, user access to an object (that is, a file or folder) is controlled by the owner of the object. The owner of an object is usually the creator of the object unless ownership rights have been taken by an administrator or supervisor (ultimate rights) and granted to another user. The Bell-LaPadula model is based on this access control.

- √ The principle of least privilege is a theory that says every user should be granted the very minimal level of permission required to perform their job properly.

- √ In a nutshell, *centralized access control* means that control over rights, permissions, user IDs and system policies are maintained in one company location on one computer system.

- √ With decentralized access control, control of rights, permissions, user IDs and system policies can be managed from several company locations using many computer systems. The Windows NT domain models are based on this concept.

- √ *Mandatory Access Control (MAC)* is a *non-discretionary* access control technique. This technique assigns hierarchical, multi-level sensitivity *labels* to users and data (as in the military—unclassified, confidential, secret, top secret). In this labeling system, user labels are referred to as security clearances while object labels are referred to as security classifications.

- √ *Role-Based Access Control (RBAC)* is essentially a type of MAC. Applications employing RBAC provide a mechanism of formulating the system's structure to compliment the existing structure of an organization.

√ Devices such as *tokens* and *smart cards* are used for authentication purposes.

√ *MAC* and *DAC* are access control techniques.

√ *RADIUS* and *TACACS* are considered a centralized access control methodology.

√ *Passwords* are considered knowledge-based authentication mechanisms.

√ *Sniffers* are software programs or devices that listen to and gather network traffic. They can be used to monitor network traffic and weaknesses or can be used to steal passwords, user IDs, or credit card information. Most network snifters work well in networks that utilize broadcast techniques. They do not operate well in networks that use collision techniques. Network transmission techniques will be discussed further in Chapter 4.

√ *Password crackers* are programs that figure out easy-to-guess passwords in encrypted password lists or databases.

√ *Remote penetration* programs are software programs that use the Internet or networks as a vehicle to gain unauthorized and illegal control of a computer system or network resource.

√ *Local penetration programs* gain unauthorized illegal access to systems on which they are run.

√ *Local* DoS programs shut down the computers on which they are run.

√ *Remote* DoS programs are used on the Internet or networks as vehicles to shut down other services or computers.

√ *Vulnerability scanners* are programs that are sent out on the Internet to search for computers that might have a weakness or certain vulnerability for a specific type of attack.

√ *Network scanners* are programs that are run on networks to map out the locations of particular networks weaknesses.

√ *Multi-factor authentication* is the combined use of a password as well as a key-exchange system to provide strong authentication. This type of authentication scheme should be implemented when the use of a Single Sign-On and password or an encrypted key system alone will not be enough security. Multi-factor authentication is considered a very strong security practice. It is likely that the exam will expect you to know this concept.

√ With *mutual authentication*, a trust relationship is first established between a host and its intended recipients or clients. Second, digitally signed certificates are typically implemented in order to allow the host or server system to authenticate to the client system. Then, the recipient or client system is authenticated with the host or server. It is highly likely that the exam will expect you to understand this concept.

√ *Weak key attacks* occur on secret encrypted keys that exhibit a poor level of encryption. Please refer Chapter 5 for more information regarding weak keys. The exam is most likely to ask you about weak keys!

CHAPTER SUMMARY

From studying this chapter, you should have gained a basic understanding of security concepts in general. You should, at the very least, be familiar with general security-related terms, access control models and techniques, authentication methods, security monitoring and auditing techniques, and the most common system and network related attacks.

For a chapter titled "General Security Concepts," this chapter might have seemed quite intense. Although the CompTIA Security+ exam is described as an entry-level certification exam, you must keep in mind that the topic of security is a very broad subject that covers a plethora of different material. Some of the topics described in this book might seem trivial while others seem more complex and require detailed explanations. As proven with other available security certification exams, it is a good understanding of all the domain subject matter that will most likely get you Security+ certified.

If you are an employer, you should seriously consider staffing a full-time network security specialist. Your network administrators, Web developers, and database programmers are likely much too busy to do their own jobs properly and keep security policies, devices, and procedures in working order. Hackers, crackers, and even plain computer novices alike are working full time to figure out ways to ruin your profits. Just do it. Put security at the top of your budget priority list or you will eventually pay the price.

Next, see if you have covered the material in this chapter well enough to answer all of the review questions correctly. Warning! One too many wrong answers on the real exam can be the difference between pass and fail.

REVIEW QUESTIONS

1. **Of the following choices, which technique is used to falsify or imitate another user's identity in order to gain access to a system?**

   ○ A. Sniffing.
   ○ B. Encrypting.
   ○ C. Spoofing.
   ○ D. Decrypting

   Correct answer = C

   Spoofing is used to allow intruders the ability to act or pretend to be an authorized user. Sniffing refers to the gathering of network packets. Encrypting is a conversion process where plain text is converted to cipher text through cryptography algorithms in order to make information secret or unreadable. Decrypting is the process of taking information or data that has been encrypted and transforming or deciphering it into plain text format. Cryptography methods are used for this conversion process.

2. **Which of the following choices best describes an ACL?**

   ○ A. Uses a mathematical format to determine if access should be granted.
   ○ B. Creates audit messages when protected objects receive access requests.
   ○ C. Authentication Clarification List.
   ○ D. A table used to identify user access rights assigned to an object.

   Correct answer = D

   An ACL (Access Control List) is a list of access rights that are assigned to an object in a system. The System Access Control List (SACL) creates the audit messages when protected objects receive access requests. The Authentication Clarification List is a trick and does not exist. Lattice-Based Access Control (LBAC) is an access control method that uses a mathematical format to determine if access should be granted to a system.

3. **Which of the following choices is used to unlock a previously encrypted message?**

   ○ A. Undercover security guard.
   ○ B. Certificate Authority.
   ○ C. Private key.
   ○ D. Public key.

   Correct answer = C

   A private key is a secret part of a key pair that is used to decrypt or unlock a previously encrypted or locked message. If you picked choice A, please read Chapter 2 again. A certificate authority is a trusted provider or issuing organization that provides and guarantees digital certificates. A public key is the known part of a key pair that is used to encrypt or lock a message.

4. **Which of the following choices best defines a password?**

   ○ A. A token-based authentication mechanism.
   ○ B. A knowledge-based authentication mechanism.
   ○ C. A characteristic-based authentication mechanism.
   ○ D. A physical security authentication mechanism.

   Correct answer = B

   A password (which is a knowledge-based authentication method) is typically compared to a security database on a host. If all requirements are met, the user is allowed access. A smart card is a token-based authentication device or mechanism. Retina and fingerprint scanning are characteristic-based authentication mechanisms. Security guards, padlocks, and gates are examples of physical security.

5. **Which type of access control technique uses a mathematical formula or lattice to determine whether or not access should be granted to a resource?**

   ○ A. LBAC.
   ○ B. DAC.
   ○ C. MAC.
   ○ D. RBAC.

Correct answer = A

LBAC uses a mathematical formula, or lattice, to determine whether or not access to a resource should be granted. DAC restricts access to resources based on the identity of users and/or groups of which they are members. This technique is referred to as discretionary because it also allows users to pass on access permissions to other users within the organization. MAC assigns hierarchical, multilevel sensitivity labels to users and data (as in the military—unclassified, confidential, secret, and top secret). RBAC enables organizational planners to put an access control policy into place that the system must abide by, rather than a policy that changes with the whim of a user or administrator.

6. **Which type of access control technique is discretionary and restricts access to resources based on the identity of users and/or groups?**

   ○ A. TOBACK.
   ○ B. MAC.
   ○ C. DAC.
   ○ D. TBAC.

   Correct answer = C

   Discretionary Access Control restricts access to resources based on the identity of users and/or groups of which they are members. This technique is referred to as discretionary because it also allows users to pass on access permissions to other users within the organization. TOBACK is the last name of a contributor of this book. Mandatory Access Control, also called nondiscretionary, uses a centralized approach to restrict access to data based on the sensitivity of the data in question. TBAC bases its access decisions on the current state of works in progress.

7. **Which of the following protocols represent centralized access control? (Choose two)**

   ☐ A. MAC.
   ☐ B. LBAC.
   ☐ C. RADIUS.
   ☐ D. DAC.
   ☐ E. TACACS.

Correct answers = C and E

RADIUS is a scalable, client/server-based UDP protocol used to transfer authentication and authorization data between a dial-in client and a server. TACACS is a dated, remote authentication protocol typically implemented on UNIX servers. Both RADIUS and TACACS are **protocols** that represent centralized access control. MAC, DAC, and LBAC are not protocols; they are access control techniques and models.

8. **Which of the following choices best represents a strong password?**
   - A. brians.
   - B. An7!$4Dt.
   - C. bob1965.
   - D. 123abc.

Correct answer = B

A strong password should be a minimum of seven characters in length. The password should contain a combination of upper- and lowercase letters. There should be at least one number contained in the password. The password should contain at least one of the following characters: ! @#$%^&*. A password should never be a word from a dictionary, a person's name, family name, phone number, birth date, or favorite phrase. All choices other than B do not meet these criteria.

9. **Which of the following represent *logical* security controls? (Choose two)**
   - A. Antivirus software.
   - B. Performance assessments.
   - C. Battery backups.
   - D. Passwords.
   - E. Security training.

Correct answers = A and D

Logical controls refer to systems such as access control software, antivirus software, communications hardware (routers), passwords, and smart cards. They represent the systems in place that prevent unauthorized access to digitized information. Administrative controls are the personnel-related mechanisms for managing people's behavior. These include security training, background investigations, mandatory vacations, and performance assessments. Physical controls entail the use of instruments such as locks, fences, motion detectors, battery backups, and burglar/fire alarms. They are the material systems and devices that protect assets from theft, fire, or other means of destruction.

10. **Which of the following are considered to be Denial of Service (DoS) attack methods? (Choose three)**

   ☐ A. MAC attack.
   ☐ B. Buffer overflow attack.
   ☐ C. SYN attack.
   ☐ D. Smurf attack.
   ☐ E. Spam attack.

   Correct answers = B, C, and D

   A denial of service attack is designed to render a network or related service useless by flooding the network with worthless or useless network traffic. Buffer overflow, SYN, and Smurf attacks are all designed to meet these criteria. If you chose MAC attack, you are probably hungry and should get something to eat before you continue on with the next chapter. Spam is the proliferation or sending of unsolicited e-mail or electronic advertisements and messages.

11. **Which attack methods do crackers most commonly use to figure out passwords?**

   ○ A. Ticket and token attacks.
   ○ B. SYN and Smurf attacks.
   ○ C. Buffer overflow and spamming attacks.
   ○ D. Brute force and dictionary attacks.

   Correct answer = D

   Crackers most commonly use brute force and dictionary attack methods as tools of choice to figure out passwords. Tokens and tickets are used as characteristic-based authentication mechanisms. SYN, Smurf, and buffer overflow attacks are denial of service attack methods used to disable access to a network or service. Spamming attacks are used to send out unsolicited e-mail advertisements and messages.

12. **Which choice is an "electronic credit card" used to verify one's identity for Internet or electronic transactions?**

    ○ A. Digital certificate.

    ○ B. Token.

    ○ C. Private key.

    ○ D. Certificate authority.

    Correct answer = A

    A digital certificate is like an electronic credit card used to verify one's credentials for Internet or e-business transactions. A token is a small credit-card sized security device that contains an ever-changing identification code, which allows its holder to access network resources. A private key is the secret part of a key pair that is used to decrypt or unlock a previously encrypted or locked message. A certificate authority is a trusted provider or issuing organization that provides and guarantees digital certificates.

13. **Which choice compares the patterns of an attack against a known system footprint before reporting an intrusion?**

    ○ A. Signature intrusion analysis.

    ○ B. Statistical intrusion analysis.

    ○ C. TACACS analysis.

    ○ D. Event Viewer analysis.

    Correct answer = B

    Statistical intrusion analysis is the process of establishing a known footprint or baseline of a system's usage of such things as CPU (central processing unit) utilization, disk utilization, use of user rights, user log-ins, file and folder access over time, and analyzing the system for any deviation from the system's baseline or "normal" behavior. Signature intrusion analysis uses a method of comparing suspect system activity against a database of known attack method patterns. TACACS analysis and Event Viewer analysis are invalid choices.

14. **What does a Kerberos ticket do?**

    ○ A. Analyzes and tests the security of a computing system.

    ○ B. Allows users to access physically secured areas.

    ○ C. Allows users access to objects.

    ○ D. It's freeware that analyses security holes in a network

Correct answer = C

The Kerberos authentication protocol uses a string of encoded messages and the issuance of special tickets to verify the identification of the user in question and allow or disallow a user's access to objects. Non-interference is designed as a tool for analyzing or testing the security of a computing system. Kerberos has nothing to do with allowing users access to physically secured areas. A computerized or electronic door or gate pass typically provides this function. SATAN is a freeware program that analyses security holes in a network.

15. **Which security model was the first mathematical security model used to address security, modes of access, and a set of rules for assigning security access rights?**
    ○ A. Biba model.
    ○ B. Clark-Wilson model.
    ○ C. OSI model.
    ○ D. Bell-LaPadula model.
    ○ E. Wilson-Phillips model.

    Correct answer = D

    The Bell-LaPadula model was the first mathematical security model used to address security, modes of access, and a set of rules for assigning security access rights. The Biba security model was created to address some of the particular weaknesses in the Bell-LaPadula model. It was not the first model. The Clark-Wilson security model was developed in 1987. Its main focus is to protect the integrity of data through the use of secured programs. The OSI model is a widely accepted seven-layered reference model that identifies how data should flow from one location to another in a computer network. The Wilson-Phillips model is not a valid choice.

16. **Which authentication method requires a user password and a digitally signed certificate to allow access to a system or resource?**
    ○ A. Mutual.
    ○ B. Multi-factor.
    ○ C. Single Sign-On.
    ○ D. Smart card.

    Correct answer = B

Multi-factor authentication is the combined use of a password as well as a key-exchange system to provide strong authentication. This type of authentication scheme should be implemented when the use of a single sign-on and password or an encrypted key system alone will not be enough security. With mutual authentication, a trust relationship is first established between a host and its intended recipients or clients. Second, digitally signed certificates are typically implemented in order to allow the host or server system to authenticate to the client system. Single Sign-On (SSO) provides a secure way for users to be authenticated just once while enabling enterprise-wide access to data. A smart card is a token-based authentication device that allows its owner to gain access to a particular service such as banking, parking, or gasoline services.

17. **Which type of biometric device is considered to be the most secure?**

   ○ A. Fingerprint scanner.

   ○ B. Signature scanner.

   ○ C. Security dog (K9).

   ○ D. Retina scanner.

Correct answer = D

A retina scanner is currently the most secure biometric device available. Fingerprint devices and signature scanning devices are the most widely accepted forms of biometric implementation. However, they are not as secure as retina scanning devices. Although security dogs can be very faithful and effective, they are not considered to be secure biometric authentication methods.

REFERENCES

http://www.microsoft.com/technet/treeview/default.asp?url=/technet/columns/ security/essays/10imlaws.asp is a Microsoft Web site that describes the Ten Immutable Laws of Security.

http://www.sans.org/newlook/resources/glossary.htm/ is a SANS Institute Web site that offers an excellent list of security and intrusion detection related terms.

http://www.securitypanel.org/glossary.html/ is a JTM multimedia site that offers another great list of security-related terms and definitions.

http://msdn.microsoft.com/library/default.asp?url=/library/en-us/ security/Security/using_security_packages.asp is an MSDN Microsoft Web site that explains Kerberos and other Microsoft security-related technical information.

http://www.cisco.com/warp/public/44/solutions/network/security.shtml/ is a Cisco Systems Web site that describes Cisco network monitoring tools as well as other Cisco security products.

http://cs.baylor.edu/~donahoo/NIUNet/hijack.html is a Baylor University Web site that has a great description of TCP/IP connection hijacking.

CHAPTER 3

COMMUNICATIONS SECURITY

IN THIS CHAPTER

- Remote Access
- E-mail Security
- Web Security
- Directory Security
- File Transfer
- Wireless Security
- Test Tips
- Chapter Summary
- Review Questions
- References

The focus of this chapter is communications security. Technically, communications security should really be covered as a subset of network security and included in Chapter 4. However, in order to keep you in line with the current CompTIA Security+ domain objectives and exam structure, this book has been tailored to follow the CompTIA domain and domain subset structure. Therefore, communications security has its own chapter.

The first two chapters of this book introduced you to several of the topics discussed in this chapter. This chapter will further assist you with understanding the details of these topics as well as introduce you to several new topics related to network communications security.

REMOTE ACCESS

Remote access is defined as the ability to make a connection to a network or computing system from outside the normal internal network perimeter. In human terms, it is the ability to connect to a network from a distance. Typically, remote access is provided to roaming users such as telecommuters, sales, marketing, and business travelers requiring access to their company's WAN or LAN resources.

Remote access is usually acquired via a dial-up connection through a modem or a more expensive dedicated line that connects a client computer to a remote network. Modern day cable modems, DSL (Digital Subscriber Line), and wireless technologies can also be used to allow remote access at much faster speeds than the traditional dial-up analog connections.

The main benefits of implementing and using remote access technologies can be seen in the reduction of cost and support of dedicated lines and the ability of workers to extend services outside of the LAN or office area. In other words, the people who know your products and services can meet with and support clients, potential clients, and vendors. Unfortunately, remote access to a network provides major security risks and is therefore a major topic that demands attention when setting up a new network or protecting an existing one.

In Chapter 4 we will discuss networking devices such as routers, firewalls, and VPNs (Virtual Private Networks) in detail. These are several of the main methods implemented in modern networks to minimize the threat of unauthorized access to an internal network through external network connections.

A remote access server or *communications server* is typically used in combination with a router, firewall, and/or VPN at the local network level to authenticate and provide security for remote users wishing to gain access to local area network resources.

As stated earlier, you have already been introduced to several of the items discussed next. By the end of this chapter and ultimately the end of this book, the security information discussed should become second nature to you.

The following are important remote access connection technologies to remember when preparing for the exam:

- **ISDN (Integrated Service Digital Network):** Carries data and voice over traditional telephone networks. ISDN will be discussed in detail in Chapter 4.
- **DSL (Digital Subscriber Line):** Considered a better and faster replacement for ISDN, DSL is a technology that can provide considerable bandwidth capabilities to small business and homeowners alike. DSL uses traditional existing twisted pair telephone lines.
- **Cable modem:** Considered the most insecure technology based on the fact that a default installation does not provide firewall or any other sort of packet filtering. With a default installation method, users share a single coax cable connection.
- **Wireless technologies:** Fastest growing area for connectivity. Wireless technology will be discussed in detail later in the chapter.
- **Dial-up (asynchronous):** Traditional connection method that uses an ISP (Internet service provider) and an analog phone line to connect to the Internet.

The following are important remote access authentication systems and secure connection methods to remember for the Security+ exam:

- Authentication systems for remote access security:
  - RADIUS -Remote Access Dial-in User Server)
  - TACACS -Terminal Access Controller Access Control Server)
- Security authentication protocols most often used for remote nodes:
  - PAP (Password Authentication Protocol) clear text
  - CHAP (Challenge Handshake Authentication Protocol)

To secure remote access connections, use the following, which are detailed in Chapter 4:

- VPN (Virtual Private Network)
- SSH (Secure Shell)
- SSL (Secure Sockets Layer)
- Firewalls

RAS (REMOTE ACCESS SERVICE)

RAS (Remote Access Service) is provided with Windows NT, 2000, and XP that allows remote clients to access services and resources located on a network using an analog modem, WAN, or ISDN connection. In order to install and use RAS, you will need a supported protocol, such as TCP/IP or IPX/SPX, and a RAS client or compatible PPP client package.

A RAS client that has been configured properly can dial in or connect to a network that utilizes a RAS server. The RAS server is used to authenticate the remote users and allow them access to services, such as file and print, which reside on a LAN.

In general, the following events occur when one establishes a connection with a RAS server:

1. Your system attempts to access the RAS server.

2. Based on your authentication methods, the following events occur:

    a. Using PAP (Password Authentication Protocol) clear text authentication:
       - Your system transmits your clear text password to the RAS server.
       - The server compares your credentials with its security database.

    b. Using CHAP for authentication:
       - The server sends a challenge message to your system.
       - Your system replies to the server with an encrypted response.
       - The server compares the response to the credentials in its user database.

    c. Using certificate-based authentication:
       - The server requests credentials from your system and sends a server-based certificate.
       - If your system and connection are configured to validate and accept the server-based certificate, the certificate is validated. If it is not validated, this step is skipped.
       - Your system then sends its certificate to the server system.
       - The server system verifies that your sent certificate is valid and that it has not been revoked.

3. If the server verifies that the account is valid, it then searches for remote access permission.

4. If you are granted successful remote access permission, the server system establishes your session and connection.

5. If you have callback enabled, the server system will call your system and repeat Steps 2 through 4.

RAS can be configured for node authentication by enabling callback or caller ID. *Callback* occurs when a remote user dials into the server and supplies credentials. The server then hangs up the call and calls the remote user back at the same number the call came in from. Caller ID occurs when the server verifies that the incoming call is matched against a predefined phone list stored on the server. It is much more difficult to gain illegal access through caller ID. It is also much more difficult to administrate caller ID for remote users who travel often and use different phone numbers to call from.

Here are a few a rules to follow when using Remote Access Service:

- The passwords that your remote users use when dialing into a RAS server should not be the same as the passwords they use to be authenticated on a domain.
- The actual time frame in hours that remote users can dial into a RAS server should be limited to the business needs of that user.
- Always use callback security in RAS for authentication.
- Your RAS server and your remote clients should always have the latest operating system service packs and patches as well as up-to-date antivirus protection.

802.1X

802.1X is an IEEE standard for wireless connectivity that uses port-based access control. It falls under the influence of the initial IEEE standard 802.11 for Wireless Local Area Networks (WLANs). The IEEE (Institute of Electrical and Electronics Engineers) and the standards that apply to networking technologies in general are explained in Chapter 4. In this chapter, we will discuss all of the IEEE 802.11 standards for wireless networking.

The 802.1X standard is designed to provide a better framework that supports improved security for users on wireless networks by the implementation of centralized authentication. 802.1X uses the Extensible Authentication Protocol (EAP), which enables the technology to work with wireless, Ethernet, and Token Ring networks.

With 802.1X authentication, a wireless client who wishes to connect to and be authenticated on network is called a *supplicant*. The supplicant must first request access from an access point, which is also known as an *authenticator*. If the access point detects the request for access from the supplicant, the access point will enable the supplicant's port and only let 802.1X traffic be transmitted. This allows the client to transmit a start-up message known as an *EAP start message*; the supplicant's identity and credentials are then provided to the access point.

Next, the access point transmits the information to an authentication server, which is typically a server that runs RADIUS (Remote Authentication Dial-In User Service). The authentication server can use various algorithms to allow the user to be authenticated, eventually. Once the server authenticates the validity of the user, it will transmit either an acceptance or rejection acknowledgement of the client's request to the access point. If the access point receives positive feedback from the RADIUS authentication server, the access point will enable or activate the supplicant's port for normal network traffic. See Figure 3.1 for a visual regarding 802.1X authentication.

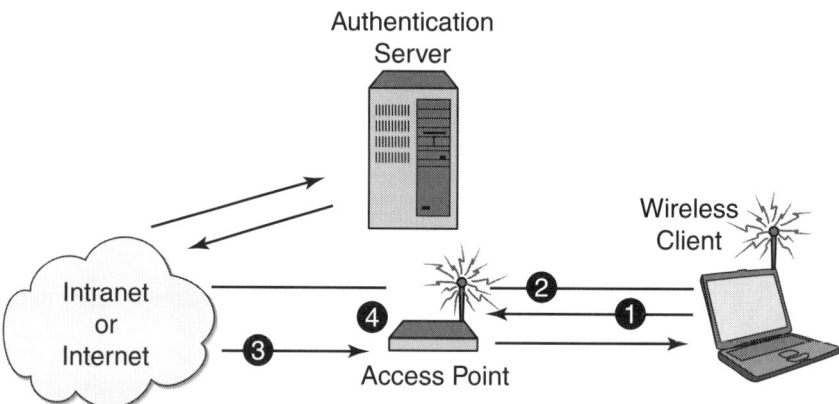

FIGURE 3.1   802.1X authentication.

To best understand this process, match the following descriptions with their corresponding numeric values in Figure 3.1:

1. A start message is sent from the remote client to the access point and the access point asks the client for identification.
2. The client sends its identity to the access point. The access then transmits or forwards the client's identity to an authentication server.
3. The authentication server transmits an accept or reject message to the access point.

4. If the access point receives an accept message from the authentication server, the client's port activates and the client is allowed to communicate with the server.

The 802.1X standard is fairly new and it is likely that CompTIA will target it on the exam specifically within a wireless or remote connectivity related question.

Microsoft does a great job explaining this technology. If you are interested in learning more about 802.1X, you may find the following site very informative: *http://www.microsoft.com/windowsxp/pro/techinfo/planning/wirelesslan/solutions.asp.*

E-MAIL SECURITY

At home or work, e-mail is arguably the most important communication tool available in today's fast paced technical world. E-mail is a fast, inexpensive, acceptable way to transfer information and communicate. Unfortunately, the use of e-mail and attachments to e-mail messages provide a very large security threat to the welfare of systems and networks in general. Providing secure e-mail and messaging systems has quickly become a top priority for home users and businesses alike.

THREATS

There are many threats associated with the use of e-mail. Virus threats are usually the most thought of but threats in the form of malicious content, leaks of confidential information, and the threat of those not-so-innocent spammers are becoming much more of a concern today. Server and workstation downtime, as well as the basic loss in human productivity that can result from these threats, are motivating the penny pinchers to invest more in e-mail and network security.

The most obvious threat to e-mail and e-mail systems comes in the form of e-mail related viruses. E-mail attachments offer an excellent opportunity for dangerous viruses to be proliferated and released even in a firewall-protected corporate network environment. *Firewalls* protect intranetworks from unauthorized user access. Unfortunately, they do not scan or filter e-mail and attachments. It is important to implement and enforce a proactive plan that includes an e-mail antivirus system with updated virus definitions that scan all attachments that are included with incoming e-mail.

It is also imperative that e-mail users are educated and informed of the dangers that can occur with the improper handling of dangerous e-mail messages. In other words, good e-mail security plans should include the empowerment of users to assume some basic common sense when opening e-mail and attachments that are suspicious.

Our basic human nature tells us to click on things we shouldn't. If curiosity killed the cat, lack of common sense killed the human's local area network. Here is an example you can use as a demonstration. Create a Hotmail or some form of Web mail account with an obscure address. Then, send an innocent e-mail with an executable attachment to five people you know. Name the subject, "Don't Open This, It's Very Dangerous." Check to see how many of the five people opened the e-mail. Get the meaning? Could this possibly be considered a form of social engineering?

Confidential information leaks from within corporate infrastructures seem to be very prevalent these days. Disgruntled employees or employees with little common sense can forward company data and personal information to the outside world. This can lead to serious legal ramifications.

Spammers, or unsolicited commercial mail distributors, can use corporate mail servers as hosts to spread their advertisements within a company and beyond. This can cause a company to be blacklisted with Internet service providers and cause a reduction to the productivity of a mail server.

Here are some basic guidelines to file that can save you or your company a lot of time, money, and sanity when it comes to protecting e-mail:

1. Develop a corporate e-mail security policy that defines a set of rules your users should follow. In other words, take the time to educate them on some of the common practices that should be followed, such as confidentiality and general security practices.

2. Install a proven antivirus protection product that scans all inbound and outbound e-mail and attachments. The product should provide an alerting service and a centralized quarantine snap-in to isolate viruses if detected. Very importantly, if your users have the ability to use outside mail services from their desktops, install a desktop solution that scans IMAP and POP3 mail services.

3. Install content-filtering software that scans for key phrases and information in e-mail that might be confidential to your company.

4. Install an anti-spam software product that will identify and allow you to block out unwanted/unsolicited information.

S/MIME (SECURE MULTIPURPOSE INTERNET MAIL EXTENSIONS)

S/MIME (Secure Multipurpose Internet Mail Extensions) is a method/protocol used to secure the sending of messages between various e-mail clients. It is based on the RSA (Rivest-Shamir-Adleman) encryption system. S/MIME consists of an updated set of standards that improve upon the original version of MIME. Its main function is to provide authenticity and privacy for e-mail messages in MIME format. S/MIME is included in popular Web browsers that are offered from such vendors as Netscape and Microsoft. S/MIME will also be covered in Chapter 5.

 It is likely that you will be asked what S/MIME is associated with. If offered the choice, and there will most likely be a choice, a good selection would be, "e-mail security."

PGP (PRETTY GOOD PRIVACY)

Developed by Philip R. Zimmerman in 1991, *PGP (Pretty Good Privacy)* is one of the most common and easiest to use methods for encrypting and decrypting messages over the Internet. It is also free.

PGP uses public key encryption based techniques to secure messaging. With PGP, a user creates a key pair and protects the private key with what is known as a *pass phrase*. The public key part of the key pair is used for the encryption of messages for other users.

 When using PGP, the pass phrase that protects the private key must not be forgotten. If the user forgets the pass phrase, the message cannot be deciphered.

In a nutshell, first, PGP will create a session key for a message that is to be encrypted. Next, PGP uses the IDEA (International Data Encryption Algorithm), which uses a block cipher or symmetric cipher and a 128-bit key to encrypt the message. Then, PGP uses the RSA algorithm to encrypt the session key with the receiver's public key. Finally, the packaged message and key are ready to be sent. PGP will be further discussed in Chapter 5.

HOAXES

A computer-related *hoax* is a myth or false representation regarding a computer-related virus. The hoax can be either a false warning or an actual file or object that closely resembles that of an actual virus. Hoaxes are usually meant to scare and cause security related hysteria. Typically, hoaxes or the news of possible fake threats spread through the Internet and corporate infrastructures like wildfire. They are like headlines or new office gossip. "Did you hear about the new eat-your-computer-alive virus?" Then, almost as quickly as the hoax is created, office staff and administrators scurry to find a fix.

The most dangerous of hoaxes will warn the recipient that there is a dangerous file and/or program residing in a system. The hoax will suggest that the recipient remove the dangerous file or program. The bad news—the recipient just wiped out an important system file that is needed for the operating system or program to exist or function. Real nice; we can only hope that the recipient has a good backup!

Hoaxes can cause a tremendous loss in productivity. The time and resources it takes to investigate hoaxes could be much better spent on actual company production issues. Unfortunately, hoaxes are part of the electronic world in which we live and must be handled with care. Could a hoax be another form of social engineering? You bet.

The best way to tell if an attachment or suspect file name is actually a hoax is through research and education. A great Internet site that allows you to do a computer virus hoax search from A-Z is *http://vmyths.com*.

If you have a good antivirus program and support contract, you can always consult the manufacturer's Web site or call them for valid viruses/hoax verification.

As a basic rule of thumb, you should always assume that a hoax or threat is real but use common sense. Don't take action based on unprofessional advice or hearsay. Finally, by no means should you apply fixes or patches without consulting a legitimate reference.

NOTE   Be prepared to answer questions that ask you how you would react to a hoax. Do you go ballistic and get everyone worried, causing large amounts of network traffic with a bombardment of e-mail and IMs (instant messages)? Or, do you remain calm and quietly investigate to see if the hoax is real? Please choose the second option.

WEB SECURITY

Web security is a huge topic that covers a wide range of security holes such as weak ActiveX controls, Java applets, HTML documents, insecure Web servers, and many other topics. In fact, we could write an entire book on this topic alone. Fortunately, our focus has been fine-tuned, thanks to the

CompTIA Security+ objectives. The topics in this section are directly in line with objectives as they relate to Web security.

SECURE SOCKETS LAYER (SSL) AND TRANSPORT LAYER SECURITY (TLS)

Secure Sockets Layer (SSL) and *Transport Layer Security (TLS)* are commonly used security protocols that provide transport security through Internet browsers offered by Netscape and Microsoft.

SSL and TLS are session-based X.509 digital certificate supporting protocols that use a public and private key exchange to encrypt the passing of data between client and server systems. Both protocols support RSA, DES, IDEA, 3DES, and MD5. They can be used by services such as HTTP, FTP, SMTP, IMAP, POP, and Telnet.

 There is a high probability that the exam will want you to know that SSL is X.509 based and uses a public and private key exchange to encrypt data.

SSL uses a combination of the SSL Record protocol and the SSL Handshake protocol to provide security. The Handshake protocol provides authentication services while the Record protocol provides for a secure connection. TLS was developed after SSL and has succeeded SSL in most uses. TLS ensures that no external or malicious forces can eavesdrop on messages sent or received.

Many Internet Web sites utilize SSL or TLS as a secure means of obtaining confidential customer information, such as bank account and other personal information. It provides confidential Web sessions and authentication services for Web servers.

Web site URLs that utilize these protocols begin with *HTTPS://*.

HYPERTEXT TRANSPORT PROTOCOL SECURE (HTTPS)

HyperText Transport Protocol Secure (HTTPS) is a very popular secure protocol used to transmit messages over the Internet. SSL and TLS are used to establish a secure connection between a client and host. HTTPS was developed and is concerned with the secure transmission of individual messages between client and host by using TPC port 443 as opposed to the normal port 80 used to transmit normal HTTP data. It is important to note that HTTPS and SSL compliment each other for secure Internet connectivity.

You will also know you are using HTTPS when you make a request through your Internet browser and the URL (Uniform Resource Locator) begins with HTTPS://.

It is probable that you will need to know that HTTPS is a secure Internet protocol for the exam.

INSTANT MESSAGING

As you are probably aware, *Instant Messaging* (IM) is a form of electronic communication service that allows a private, or not so private, conversation to take place between two or more people. You can download popular instant messaging software over the Internet from providers such as AOL, Yahoo, MSN, and ICQ. A popular XML-based, open-source instant messaging system known as Jabber can be also be downloaded. More information can be obtained from *Jabber.org*. With instant messaging, a user is typically notified when a "friend" or "buddy" is available or online. Either user can then initiate a messaging conversation as long as the possible recipient of the message is willing to accept it. Some services include the ability to block messing from those pesky unwanted users that want to chat at an inappropriate time. You can even have bells and whistles go off when someone wants to chat with you.

Instant Messaging Vulnerabilities

The instant messaging craze has swept the world by storm. What better way to communicate could there possibly be? Instant messaging is quick and convenient. It allows people at work to communicate with others at work very quickly throughout the corporate empire and also allows workers to chat with their families while on company time.

Instant messaging software is everywhere and hackers, crackers, and would-be thugs are using the security holes that instant messaging offers to do their wrongful deeds.

Most client-side instant messaging software of the day is Web based. People who seem to be hard at work in offices and cubicles are often chatting on instant messaging systems with software they have downloaded personally. These messaging systems often go through the Internet, bypassing corporate firewall security. This is a big no-no for obvious reasons.

Many messaging systems allow users to send attachments and do file transfers of information. Although many companies and home users have antivirus programs that scan e-mail and attachments included with e-mail, little protection exists for messaging systems.

Another popular hacking technique is to attack a home user through an instant messaging system at home sweet home, then attack the company the home user works for through the user's remote VPN connection to the office. More and more buffer overflow attack methods are being instituted through the use of popular instant messaging programs.

The following are recommendations that should be implemented to protect your home or office network from the threats of instant messaging vulnerabilities:

1. Install a personal firewall at the desktop level.

2. Purchase an intra-company messaging system that includes specialized secure IM servers and client software. Yes, some secure technologies will cost money.

3. If possible, disable or eliminate the ability of the messaging client to carry out file transfers.

 There is a huge chance that you will have to identify the security weaknesses involved with using instant messaging. My guess would be that the number-one weakness of instant messaging is that it was developed without concern for security. Number two weakness: messages and transfers are not encrypted. Clear text all the way!

8.3 NAMING CONVENTIONS

Understanding 8.3 naming conventions is important for any network technician or administrator. Although 8.3 naming conventions are part of the CompTIA Security+ domain structure, it is not probable that you will have to do name conversions or identify the improper use of names. However, for the domain structure study and just in case, know this material.

DOS files are stored in directories or subdirectories. In today's Windows world, we call directories *folders*. In DOS, there are specific rules that apply when creating and naming files and directories.

DOS uses what is called an 8.3 file naming structure, otherwise known as *eight dot three*. This simply means that a file name can be up to eight characters long and have a three-character extension that represents the file type. A period (.) or *dot* is used to separate the file name from the extension. This means that the total length of the entire DOS file name plus the extension cannot exceed 11 characters. The file extension is not necessary unless the file is associated with a particular function.

Let's use the DOS file name AUTOEXEC.BAT as an example. The *AUTOEXEC* is the DOS file name. The *BAT* extension specifies that the file is a batch file. The same is true for the CONFIG.SYS file. The *CONFIG* is the name of the file. The *SYS* extension identifies the file as a system file.

The following are rules that apply to DOS file and directory name creation:

- A file or directory name can be no more than eight characters long.
- An extension can be no more that three characters long.
- No spaces can be included in the file, extension, or directory name.
- The characters (? * , ; = + < > | [ ] / \) are illegal and cannot be used.

Long File Names (LFNs)

Windows 9.*x*, Windows 2000, Windows XP, Macintosh systems, and UNIX all support Long File Names (LFNs). LFNs can generally be up to 255 characters in length. Although these newer operating systems support LFNs, they still allow for backward compatibility with the 8.3 naming structure associated with DOS by creating an associated 8.3 file name for every new file created. Long File Names are basically broken down into 12-byte sections that allow for up to 255 characters to be utilized. For example:

- The Long File Name=*BEST CERTIFICATIONBOOK.DOC.*
- 8.3 associated filename=*BESTCE~1.DOC.*

Notice that the space was eliminated in the 8.3 associated file name. Windows automatically removes any spaces or invalid characters and truncates the filename.

It is important to note that the Windows 9.*x* root directory (C:\) can hold only 255 files. The truncation of LFNs to 8.3 names can fill up this 255-file storage limitation quickly and cause your system to halt. For this reason and others it is good practice to avoid storing your files in the root directory of any operating system. As a general rule of thumb, you should use all lowercase letters when naming your files and folders. Case in point, UNIX systems see upper- and lowercase files as different file names. For example, the UNIX OS will see the files *Security.txt* and *security.txt* as two separate files. You should also use intuitive, recognizable file and folder names. Meaningless names can result in files and folders that are very difficult to locate on a network.

LFN Security Precautions

Certain e-mail packages are susceptible to long file name vulnerabilities. When file attachments or news messages that exceed a certain number of characters in length are opened, the e-mail package shuts down. This inherent weakness of some mail packages is typically resolved with an upgrade or vender service patch.

Some earlier versions of Windows contain built-in networking software programs that process strings of file names. If these networking packages are not patched and receive exceptionally long file names strings, the system could experience a buffer overrun and crash.

 Exceptionally long file names are often intentionally used to exploit network software vulnerabilities and cause buffer overruns.

Universal Naming Convention (UNC)

A *UNC* name is used to access a particular share on a particular workstation or server on a network. If you want to access a resource such as a printer or folder that has been shared on the network, you can gain access to it by typing in the UNC name from the Start>Run option in Windows. A UNC name always follows the format *Servername\sharename*. For example, say you want to access a folder on your network named Certified. The Certified folder resides on a server named Bigserver. You would go to Start>Run and type in *Bigserver\Certified*. If you have not been granted access permissions to the resource, you will be asked to provide a valid user ID and password in order to gain access.

PRIVACY

Computer information privacy is a very hot topic today. Computers and the Internet have made such things as paying bills, making purchases, and gaining information on just about any topic very easy. With a simple trip from the couch to the computer, you can make your mortgage payment, find a mate, buy a car, or take a virtual trip through the castles of Germany among other things.

But what price do we have to pay for this ease of accessible information? What really happens behind the scenes when we innocently view and educate ourselves using Web pages on the Internet?

When you view pages on the Internet, your Internet browser's IP address and information such as your operating system's version and other personal information are transmitted to the server that provides the page. This is done when your browser assembles a page and makes an HTTP request to a server. That server also checks to see if a *cookie*, which is a unique identifier that Web servers place in your system to identify you, has been placed in your system. If it has not, the server gives you a cookie. As a result, every mouse click you make on that particular site lets the site owner know it's you.

Yes, you're being watched. You are also being target marketed. Personal information such as your e-mail address is being sold and/or freely distributed by companies that house huge servers and databases that distribute cookies for many of the popular Internet sites. Wonder why you get so much spam? Click on a few more of those advertisements on a site and your information will be recorded and sold again.

Although it is unlikely that you will ever be able to protect your privacy totally on the Internet, there are some important actions you can take to minimize the amount of information you provide to servers, marketers, and spammers while surfing the Internet.

Purchase a software package that will remove unwanted cookies, banner ads, and sensitive HTTP header information. There are a slew of free proxy server software packages as well as other software packages that will hide your IP address while surfing the Internet. You can also disable Java, JavaScript, and ActiveX, which provide privacy invading security holes in your system. These will all be described next.

 You might be asked to identify privacy in a manner similar to the following question: Darren wants to send Chris an e-mail that contains sensitive financial information. Darren is concerned that someone other than Chris will intercept this e-mail and gain access to the sensitive data. What is Darren concerned with?

  ○ A. Non-repudiation.
  ○ B. Authentication.
  ○ C. Encryption.
  ○ D. Privacy.

The correct answer to this question is D. Darren is concerned with privacy. Although answers A, B, and C look very tempting, they are incorrect to technical test takers. This is exactly what you need to be aware of when answering common sense questions. Be careful!

COOKIES

As mentioned earlier in the chapter, a *cookie* is a message that is sent or transmitted to a Web server from a Web browser. It is important to note that a cookie can also be referred to as a *state object* or *persistent cookie*. The cookie is used to provide the Web server with unique information that is used to identify where the request to the server is coming from. In other words, the cookie provides information about you.

When you access pages on the Internet and enter your personal information into Web forms, that information along with other prepared information is combined into a file called a cookie. Most often, server-side scripts known as *CGI scripts* are used to control what happens with the cookie. The cookie is typically placed in the user profile folder of the user currently signed on to the system. If you are using a newer version of Windows such as Windows 2000, this location will be *C:\Documents and Settings\Profile Name\Cookies*. In the author's case, this location is *C:\Documents and Settings\Drvankman\Cookies*.

The next time you happen upon the same Web site, your Internet browser will automatically forward the stored cookie to that Web sites server. If set up to do so, the server will provide you with a prepacked, customized page that targets you for specific advertisements. In other words, you will get spammed.

Many Web servers use trusted cookies as their only form of authentication. This widespread misuse of cookies has spawned a major security threat to both Web servers and users alike. If an attacker or cookie hijacker is able to infiltrate a user's session while the user is logged on to a server service, the attacker can steal the user's cookie and use it to access such things as account information. A common practice that attackers use to grab cookies during an active session is to execute a fake Java script on an unprotected server.

Although many servers are open to this cookie authentication exploitation, most servers that provide important financial information and extremely sensitive data have more secure authentication mechanisms and devices in place.

There are several good ways to protect your system from the threats to security provided so nicely by the use of cookies: You can set up you Internet browser to alert you when a cookie is present; you can direct your browser to download only cookies from trusted sites; or you can disable cookies altogether. Depending on your operating system and Internet browser, this procedure will vary.

For a great explanation of persistent cookies visit *http://wp.netscape.com/newsref/std/cookie_spec.html*.

 There is a very good chance the exam will want you to know that cookies are extremely dangerous and can carry malicious viruses.

JAVASCRIPT

JavaScript is a programming script language that is supported by Internet browsers provided by Netscape and Microsoft. JavaScript is used commonly by Web developers to interact with Web pages that are typically created with HTML or XML source codes. In simple terms, Java allows developers to spruce up new or existing Web pages by adding features such as self-updating software packages, pop-up windows, link-to pages, and 3D interactive worlds. Java is considered to be a portable, object-oriented, robust, and secure scripting language.

The productive tools that we manufacture to design and create a better, more intuitive Internet experience all seem to come with a heavy price tag concerning security. Java, JavaScript, and Java *applets* (little programs sent with Web pages that do not require user interaction) are no exception. They all provide transportation mechanisms and can allow attackers to insert code that can infiltrate and destroy your system. Java scripts and applets are programs created that actually run on your system.

Hijackers and attackers often create scripts and applets, which are oftentimes able to circumvent network security parameters. They can be used to manipulate files on users' computers.

SIGNED APPLETS

As mentioned earlier, *applets* are small programs that contain scripts sent with Web pages to users. Applets such as Java applets allow calculations, animations, and other functions to take place on a user's system without a need for communication to occur back to the applet-providing server.

Applets can contain malicious code that can easily destroy a system if allowed to run. A popular technique known as *sandboxing* is often used to quarantine applets that appear suspicious or malicious.

Signed applets contain a digital signature. They are a means of proving that an applet has come from a trusted location, author, or site. Signed applets receive permissions to access local system resources. Plain applets only have access to the directories from which they originally run. Some books state that signed applets cannot be altered. This is simply not true. Anyone can create or forge a signed applet. This makes them very dangerous and provides a huge security vulnerability to local as well as networked systems.

It is important to note that most applets on the Web are *unsigned applets*. These applets can be assigned various security levels, which include untrusted, high, medium, and low levels of security. Please refer to Table 3.1 for the various levels of unsigned applet security.

The following site is an excellent resource that describes applets in more detail than the scope of this book will permit: *http://java.sun.com/sfaq/ #prevent*.

TABLE 3.1   Applet Security Levels

Applet Security Level	Action Taken Explanation
Untrusted	*Untrusted applets* do not have permission to run on a system at all. They only have the ability to start.
High	Applets with this security rating run under what are considered *safe restraints* and are only permitted certain functions. They are not permitted to carry out unsafe actions. They cannot access most browser settings. They cannot read, write, delete, or change files. They can only listen to network ports located above port 1024. They cannot access a system's printer queue or clipboard.
Medium	These applets can run under safe restraints. By default, if one of these applets attempts to read, write, or change any of the high restraints, you will be prompted (warned) by your Internet browser. Next, you may grant the requested permission to the applet if you choose.
Low	This level carries the greatest security risk. Applets with low security run under minimum restraint. Your browser will not warn you of such actions unless the applet attempts to start local applications.

ACTIVEX

ActiveX is a set of object-oriented programs, technologies, and tools that are Microsoft's answer to Java technology, which by the way was created by Sun Microsystems. ActiveX is basically a combination or outgrowth of the Microsoft technologies known as *OLE (Object Linking and Embedding)* and *COM (Component Object Model)*. When this technology is used in a networked environment that provides directory support and other service, the COM technology becomes *DCOM (Distributed Component Object model)*.

Using this technology, the goal is to create a self-sufficient program known as a *component* or *ActiveX control* that can be run anywhere your ActiveX network exists.

ActiveX controls or components are compared to Java applets and can be reused by applications and other systems throughout your network.

ActiveX provides a power tool for developers and programmers. Unfortunately, ActiveX carries with it security risks as do Java scripts and applets. However, the ActiveX security model is quite different from the security controls in place for Java and Java applets. As you might recall, Java applets are restricted based on a set of actions that are considered safe. The ActiveX security model does not limit an application package to a set of individual restrictive controls. Instead, its controls are based on digital signatures. These digital signatures are registered and certified with a trusted Digital Authority such as VeriSign. When a person registers a software package or application with a trusted CA (Certificate Authority), they are agreeing that the package or ActiveX control is free of malicious code. From that point on, the risks involved with downloading the controls are totally the responsibility of the user.

In simple terms, the main weaknesses or problems associated with ActiveX controls are as follows:

- Once the user has accepted the certificate, responsibility of the control's actions is placed completely on the user. If an uneducated user on your network happens to accept a certificate from an unknown or unofficial CA, you might not have many systems left by the end of the day.
- Users can change browser settings to allow unsigned ActiveX controls to be downloaded with a warning.
- There is no good logging or audit trail available to track down what an ActiveX control has done to your system.

If you need help troubleshooting problems associated with active content such as JavaScript and ActiveX using Internet Explorer, Microsoft provides an excellent white paper on the subject. You will also be shown how to disable dangerous active content all together in this white paper found at *http://support.microsoft.com/default.aspx?scid=KB;EN-US;Q154036*.

CGI

The *Common Gateway Interface (CGI)* is a language-independent interface or standard that Internet Web servers use to pass a user's request to an application program and forward a response back to the Web server, which in turn provides the results to the user. In English, when a user fills out a HTML form on a Web page, typically, a CGI program is used to process the forms data behind the scenes and get the information back to server. This allows Web servers to serve and interact with the users dynamically. The actual method of passing data between a server and an application is called the CGI.

CGI programs run on Web servers and are considered to be *server-side* applications. Java scripts, applets, and ActiveX controls are run on individual systems and are considered *client-side* programs. A disadvantage with using CGI programs is that they start a new service on a Web server every time a CGI program runs. This can result in a major decrease in performance of a Web server.

The use of CGI programs allows the vulnerabilities associated with HTTP to be exploited. Also, in order for CGI programs to work, they are written to run on most operating systems and have access to important server system files as well as connected hosts.

Poorly executed CGI scripts and lack of or improper file system permissions can open the security hole doors and leave servers as well as systems vulnerable to attacks.

SMTP RELAY

SMTP (Simple Mail Transfer Protocol) is a TCP/IP protocol that is used for the sending and receiving of e-mail between servers or clients and servers. In most cases, SMTP is used for sending mail only. This is based on its inability to handle message queuing properly at the mail-receiving end. In order to receive stored messages properly from a mail server, most client-side systems are set up for *POP3 (Post Office Protocol)* or *IMAP (Internet Message Access Protocol)*. You send mail with SMTP. You download or receive mail with POP3 or IMAP.

SMTP is most often used with TCP port 25. TCP and UDP ports are described in Chapter 4.

SMTP relay occurs when an intermediary mail server (relay server) is used to accept any incoming mail and forward it to another mail server or final destination. This final destination is typically the e-mail server where the user's e-mail account is stored. There can be many relay servers involved with the relaying of e-mail.

The problem here is that mail servers implementing SMTP relay usually accept most mail received and deliver or relay outgoing mail without verifying or authenticating the sender or receiver. Spammers, spoofers, and unauthorized users can take advantage of this vulnerability by faking a sender's address and using just about any receiving address they wish. This can cause great proliferation of junk mail or spam and usually does.

A common problem among companies that use improperly configured e-mail and SMTP relay servers is that they are unknowingly being used as hosts to spam other servers and hosts. This can result in a company's mail server being *black holed*, meaning they are banned or blocked from using e-mail services provided by ISPs. This can then result in a major loss of productivity and downtime. SMTP relay anyone?

If you want to learn more regarding Web security, visit the following site that is hosted by the World Wide Web Consortium. You could probably spend the rest of your IT security career at this awesome site: *http://www.w3.org/Security/Faq/*.

DIRECTORY SECURITY

A problem for many organizations today is the ability to utilize, manage, and secure productively heterogeneous network environments that have many users and various resources. An unorganized network with mismanaged users, permissions, access controls, and scattered resources can cost a company a small fortune as well result in a major lack of productivity.

The use of directories and directory services such as Microsoft's Active Directory and *LDAP (Lightweight Directory Access Protocol)* help bring together various heterogeneous systems and networks. These services have provided a more intuitive way to manage, access, and control resources at the enterprise level. Proper implementations of directory services can improve security and reduce operating system maintenance and administrative support costs. However, lack of or improper management of directory services can prove fatal to a network and its resources.

Directory services such as Active Directory and LDAP use hierarchical tree structures as a means of organization and a way to administrate at various levels of an organization.

LDAP

LDAP is a directory service standard protocol, which is part of X.500 that is used on the Internet and many corporate intranets. LDAP allows resources such as files, folders, devices, locations, and people to be located easily on a network. On traditional TCP/IP networks, the DNS (Domain Name Service) is the most commonly used directory services protocol. DNS is used to translate or resolve domain names to specific IP addresses. For example, DNS will resolve the name CompTIA.com to the IP address of 216.119.103.72. It is easier for people to remember intuitive names than numbers such as IP addresses. LDAP allows you to search by more intuitive names in case you cannot remember or do not know the domain a user or resource is located in.

Although LDAP provides many benefits, it also has security-related vulnerabilities. LDAP is used to allow productive interoperability between X.500 supported directory services and directories. Directories typically include such things as ACLs (Access Controls Lists), certificates, and other important information. LDAP is given access to most of these resources on an enterprise scale. If any of these individual resources become compromised, your entire enterprise might be at risk.

 Known vulnerabilities exist with certain versions of LDAP that have led to buffer overflow attacks, unauthorized access conditions, and Denial of Service(DoS). It is important for you to review your version of LDAP and apply patches or upgrade to a more secure version if required.

Most major vendors including Microsoft, Novell, and Cisco all have products that interoperate with LDAP. Microsoft, for example, provides support for LDAP with its trademarked directory service known as Active Directory.

In conclusion to our Web security section, it is advised that you visit the following site for a very informative list of the top ten most exploited Internet security flaws: *http://www.orthus.com/ttvuln.html.*

FILE TRANSFER

File transfer, or to be more specific for our study purposes *FTP (File Transfer Protocol)*, is defined as the placement or movement of one or a group of electronic files from one system to another. FTP is an application layer protocol that uses TCP/IP protocols. Users can download files easily from local FTP servers or download files and programs from FTP sites on the Internet to their local systems. FTP commands can be executed from a command line or a GUI (Graphical User Interface) can be used to carry out FTP functions. Users can use FTP commands to rename, delete, copy, or move files that are located on an FTP server. More secure FTP servers require a specific user name and password in order to access files and folders.

Data that is publicly available to users can usually be accessed using *anonymous FTP*. Anonymous FTP allows users to access information on FTP servers without authenticating with a unique identification. Users can use anonymous FTP to access the FTP site by entering *anonymous* as a user ID and any password the user chooses.

As you will soon read, there are many security related issues and concerns involved with using FTP. Stay sharp here; the exam is sure to target the inherent security weaknesses associated with FTP.

S/FTP

There are many third-party programs available today that assist in making your FTP server more secure. Many of these programs are Java based and allow for an SSL\TLS encrypted connection to your FTP server. For very secure FTP, X.509 certificates and the use of *asymmetric public key cryptography* is often used to encrypt public keys. For larger file transmissions, *symmetric keys* are used to encrypt and decrypt data sessions.

There are many algorithms that are used in securing FTP. A few of these include DES, 3DES, and Blowfish. In conclusion, S/FTP (Secure/FTP) is meant to provide strong authentication and encryption services and support for FTP. Please note that algorithms, certificates, and keys will be described in detail in Chapter 5.

TFTP

TFTP (Trivial File Transfer Protocol) is a scaled down, simplistic version of FTP. Instead of using the *TCP (Transmission Control Protocol)* that FTP uses, TFTP uses the UDP (User Datagram Protocol). Unlike FTP, TFTP does not use authentication and doesn't provide any security features whatsoever. It is commonly used by servers as a mechanism to reboot diskless systems and X-terminals.

TFTP uses UDP port 69. Many TFTP servers are targeted by buffer overflow attacks that typically result in a system rest. This results in a denial of service. If you have had the wonderful experience of combating the Nimda virus, you are probably aware that Nimda scans ports 69 and 80 as a means to spread itself. If you do not need to use TFTP, it is suggested that you block traffic on port 69. You will save yourself a world of hurt and possibly many long nights at work. The Nimda virus will be explained further in Chapter 9.

It is common knowledge in the industry that unrestricted TFTP servers can be exploited remotely. Just about anyone can gain access to sensitive material if it resides on a TFTP server. If you absolutely must use a TFTP server, make certain that you have applied all operating system and vendor patches. Otherwise, disable TFTP altogether.

VULNERABILITIES

Although FTP is great tool for transferring files, folders, and programs that are too large for e-mail, there are serious vulnerabilities involved with the use of FTP. Some of the highly visible weaknesses inherent with FTP include the following:

- **FTP sessions by default are not encrypted:** Usernames and passwords are transmitted in clear text. FTP user IDs and passwords can be grabbed easily with a sniffer.
- **Unsecured FTP sessions are highly susceptible to Man-in-the-Middle attacks:** In other words, the files being transferred are grabbed, modified, and forwarded to the FTP server.
- **Port scanning attacks:** When a client connects to an FTP server and requests services, the port through which the client made the request stays open until the server responds. There are inherent problems with open ports in FTP. Attackers can run port-scanning techniques to infiltrate open ports and connections on weak FTP servers.
- **IIS (Internet Information Server):** This is Microsoft's Web server product. By default, FTP services are installed with a default IIS instillation. Novice administrators and unknowing rookies are oftentimes unaware of this fact. They leave unneeded services running on their Web servers and leave the FTP service unsecured.

Follow these good FTP practices:

- Configure your FTP server to run FTP services only. Do not run unnecessary services that can be exploited.
- Do not store valuable data that cannot be recovered on your FTP server.
- Use a secure file transfer package such as SSL and encrypt all-important data.
- Disallow unnecessary access to your FTP server. Do not use *blind* or anonymous FTP.
- Audit and log events on your FTP server.

FILE SHARING

The importance of sharing locally stored files, folders, and other shares is quite obvious in today's fast paced mobile computing world. Unfortunately, misconfiguration and the lack of proper file system administration opens security holes to an operating system and allows systems to act as catalysts to spread viruses through networks.

An inherent weakness with earlier Windows versions can be seen when file and printer sharing is first enabled. By default, all shares and connections on the system are available to anyone who can access the system locally or remotely. This can be rectified by password protecting each share. Have fun! Hope you remember all of those passwords.

If you have a small peer-to-peer network that you want to protect from outside forces and still be able to share files internally, it is advised that you do not use TCP/IP for file and print services. You can still use file and printer sharing internally if NetBIOS is running. Windows uses NetBIOS locally to communicate with other systems on the same network.

For your protection, it is highly recommended that you use a hardware firewall if you must enable file sharing on your network. If you're running a small network at home and you can't afford a good router/hardware firewall solution, download a free software firewall program. It's better than nothing.

WIRELESS SECURITY

Wireless transmission is defined as the sending of signals over electromagnetic radio waves. Wireless networks have become very widespread. In some cases, wireless networks replace the need from tradition wiring. However, in larger networks, wireless technology is used typically as an extension or addition to a wired network. Wireless networks offer the ability of computing in places that would be otherwise hard to reach with a wire or cable. The use of wireless technology has been widely accepted by the military, hospitals, business, museums, and home users alike.

The IEEE (Institute of Electrical and Electronics Engineers) has developed standards for wireless technologies. These standards are a set of rules that provide a sort of instruction map of guidelines for technology developers to follow when creating new or adding to existing technologies. The IEEE standards that apply to wireless networking are 802.11, 802.11a, 802.11b, and 802.11g. The IEEE and more information on wireless standards will be discussed in Chapter 4. In its simplest of forms a wireless local area network or WLAN is displayed in Figure 3.2. Basic wirelesses networks typically have a wireless client, an authentication server or host, and an access point.

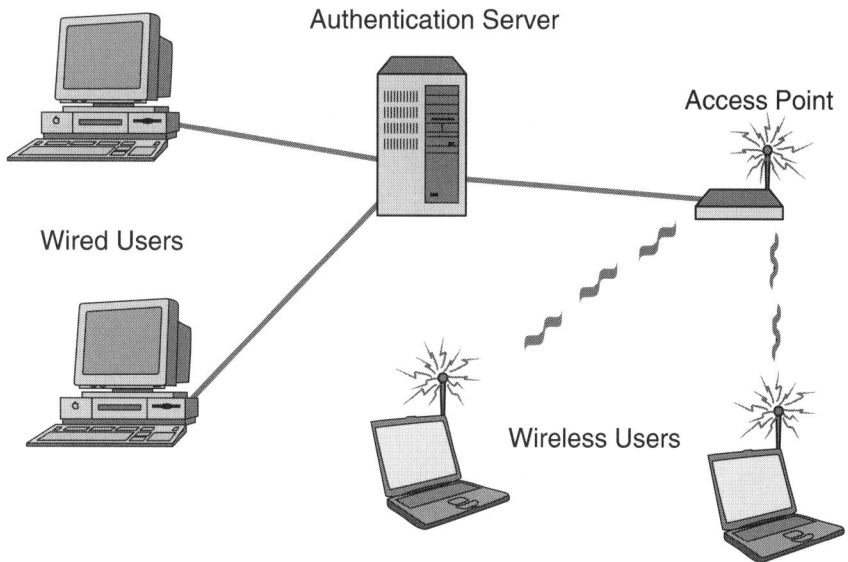

Authentication Server

Access Point

Wired Users

Wireless Users

FIGURE 3.2   A basic wireless network.

The need to secure the use of wireless technologies and remote wireless user access has become paramount. Not too long ago, wireless connections were thought to be somewhat secure. Recently, these thoughts have changed based on security holes found in the technology. Next, we will discuss several important wireless protocols that are used to secure wireless communication.

WAP

WAP (Wireless Application Protocol) is a wireless standard that applies to wireless communication protocols and devices. There are several wireless standards that are used by various wireless device service manufactures. WAP is positioned to allow interoperability between them.

WAP has its own built-in security. It uses *WTLS (Wireless Transport Layer Security)*, which is the wireless version or relative of SSL. As you might recall, SSL uses certificates and a client/server, verification/authentication process. So is the case with WTLS.

WEP

WEP (Wired Equitant Privacy) is a wireless security protocol specified under the IEEE 802.11b. WEP is intended to provide a WLAN with a similar security level as the protection that can be found in traditional *LANs (local area networks)*. WEP attempts to secure the obvious security hole that exists between a wireless client and an access point by encrypting the data that is transmitted. Once the data has been safely transmitted, it is thought that conventional network security measures such as VLANS, antivirus, tunneling, and authentication solutions can be implemented for security purposes.

 The exam is likely to ask you basic questions relating to wireless networking. For example: What technology does WEP have to do with? Or what does WAP do?

SITE SURVEYS

Before you install a wireless network solution into an existing building or between existing buildings (building to building), you should first have a professional site survey conducted by certified RF (radio frequency) engineers. These engineers can properly recommend and assist you with an integration plan as well as keep you in line with federal, state, and local regulations as they apply to wireless networks.

With traditional network installations, it is much easier to plan out a network topology and possibly foresee obstacles that will need to be addressed. However, with wireless networks that implement the use of radio transmission techniques, it is very difficult to plan for and determine how a network will react to the surrounding conditions. Obstacles such as asbestos lined walls, trees, and other physical impediments can severely impact the effectiveness of wireless communication. The interference with other RF bands in busy airways can severely hamper your performance and ability to communicate between access points. Certified site survey technicians can detect potential interference between RF bands with a tool called a *spectrum analyzer*.

A good site survey should provide you with the most suitable wireless equipment options to integrate with your current topology and applications. It should also provide you with a wireless standard that is in line with your required transmission speeds and ultimately, your budget.

 If you are interested in getting a wireless network solution, the first step is to have a site survey performed.

TEST TIPS

Here are your Test Tips for the communications security chapter. It is important to remember that some of these tips are not necessarily covered in the chapter but know them well. There is a good possibility that these tips will help you in a pinch on the real exam. Also, the tips at the end of each chapter are prepared with a special focus towards the CompTIA Security+ exam. Keep in mind, these tips will prove very useful as study for any security related exam available in the technical arena.

- √ *Hypertext Transport Protocol Secure (HTTPS)* is a very popular secure protocol used to transmit messages over the Internet.
- √ Always configure your e-mail server to block or remove e-mail that contains file attachments that are commonly used to spread viruses such as *.vbs*, *.bat*, *.exe*, *.pif*, and *.scr* files. If you must use any of these extensions, make it a point to have your e-mail antivirus program scan these attachments before delivery.
- √ Instant messaging vulnerabilities have become a popular target for modern day hackers.
- √ A UNC name always follows the format *Servername**sharename*.
- √ A *cookie* can also be referred to as a *state object* or *persistent cookie*.
- √ You can set up your Internet browser to alert you when a cookie is present; you can direct your browser to only download cookies from only trusted sites; or you can disable cookies altogether.
- √ *JavaScript* is used commonly by Web developers to interact with Web pages that are typically created using HTML or XML source codes.
- √ Hijackers and attackers often create or intercept Java scripts and applets, which are oftentimes able to circumvent network security perimeters and use them to manipulate files on users' computers.
- √ A popular technique known as *sandboxing* is often used to quarantine applets that appear suspicious or malicious.
- √ The ActiveX security model does not limit an application package to a set of individual restrictive controls. Instead, its controls are based on digital signatures.
- √ *S/MIME (Secure Multi-Purpose Internet Mail Extensions)* is a method/ protocol used to secure the sending of messages between various e-mail clients.

√ SSL and TLS are session-based X.509 digital certificate supporting protocols that use a public and private key exchange to encrypt the passing of data between client and server systems. Both protocols support RSA, DES, IDEA, 3DES, and MD5.

√ You send with mail SMTP. You download or receive mail with POP3 or IMAP. SMTP is most often used with TCP port 25.

√ FTP sessions by default are not encrypted. Usernames and passwords are transmitted in clear text. FTP user IDs and passwords can be grabbed easily with a sniffer.

√ Known vulnerabilities exist with certain versions of LDAP that have led to buffer overflow attacks, unauthorized access conditions, and Denial of Service.

CHAPTER SUMMARY

Implementing private communication channels and securing the tools we used to communicate with each other is quickly becoming the number-one priority of businesses that wish to compete and simply stay open for business.

From your study of this chapter, at the very least, you should have a basic understanding of the following:

- The tools and processes used in securing remote connections.
- The common security risks and threats associated with e-mail and Web services. You should also know the best practices to implement to safeguard these services.
- Wireless networking concepts, protocols, and security.
- The concepts of file transfer and the best security practices for securing the transfer of information.

After practicing your knowledge of communication security concepts with the review questions that follow, be sure to supplement your knowledge with the resources provided at the end of the chapter in the "References" section. The references provided at the end of every chapter in this book offer very detailed information on the particular topics of study. Use them!

REVIEW QUESTIONS

1. **Which protocol is used to transfer files and directories between two systems?**

   ○ A. Simple Mail Transfer Protocol (SMTP).

   ○ B. File Transfer Protocol (FTP).

   ○ C. Telnet.

   ○ D. Simple Network Management Protocol (SNMP).

   Correct answer = B

   File Transfer Protocol (FTP) is an application-layer protocol used to transfer riles, folders, or Web pages between two systems on a network. Simple Mail Transport Protocol (SMTP) is a TCP/IP protocol used in the transferring and receiving of e-mail messages. Telnet is a TCP/IP protocol or command used for accessing remote computing systems. Simple Network Management Protocol (SNMP) is a protocol used to manage and monitor network-related devices.

2. **What does S/MIME do?**

   ○ A. It synchronizes messages in MIME format.

   ○ B. It provides protection for e-mail messages in MIME format.

   ○ C. It replaces the need for e-mail antivirus programs.

   ○ D. It provides the ability for a two-way encrypted key exchange.

   Correct answer = B

   S/MIME (Secure Multipurpose Internet Mail Extensions) is provided in most modern day browsers. It is based on the Rivest-Shamir-Adleman encryption system and provides a secure method for exchanging e-mail messages. All other choices are invalid.

3. **Using PGP, what does the user, or creator, protect with a pass phrase?**

   ○ A. Public key.

   ○ B. Private key.

   ○ C. IDEA.

   ○ D. Key block.

   Correct answer = B

With PGP, a user creates a key pair and protects the private key with what is known as a pass phrase. The public key part of the key pair is used for the encryption of messages for other users. PGP uses the IDEA (International Data Encryption Algorithm), which uses a block cipher or symmetric cipher and a 128-bit key to encrypt the message. Key block is an invalid selection.

4. **Which of the following are true regarding computer-related hoaxes? (Choose two)**

   □ A. A hoax is a virus.
   □ B. You should always follow all instructions provided with any hoax.
   □ C. A false virus warning.
   □ D. Never take a possible hoax seriously.
   □ E. A form of social engineering.

   Correct answers = C and E

   A computer-related hoax is a myth or false representation regarding a computer-related virus. It is a false warning. A hoax is also a commonly used form of social engineering used for nonproductive purposes. A hoax can resemble a virus and many times hoaxes have names similar to viruses. However, a hoax is not a virus. You do not follow the instructions provided by a hoax until you have done your research and verified its validity. Until you know it is really a just a hoax, you should always take the hoax and other threats seriously. Just remember to do your research before you act!

5. **What tool should be used during a wireless network site survey to detect possible interference in RF bands?**

   ○ A. Packet sniffer.
   ○ B. Network monitor.
   ○ C. Spectrum analyzer.
   ○ D. Performance Monitor.

   Correct answer = C

   Certified site survey technicians can detect potential interference between RF bands with a tool called a spectrum analyzer. A packet sniffer is a program or device that collects and monitors data packets on a network. A network monitor keeps track of specific events that occur on a network. A network monitor has the ability to can produce reports and provide alerts to network administrators when critical network errors occur.

Performance Monitor is a performance-management tool included with Windows NT. It allows administrators to measure the system performance of such items as memory, CPU, and hard drive utilization.

6. **Secure Sockets Layer (SSL) is a security protocol made up of two protocols. The SSL Record Protocol and the**

  ○ A. Extensible Authentication Protocol.
  ○ B. Secure Sockets Layer Handshake protocol.
  ○ C. Extensible Handshake Sockets protocol.
  ○ D. Secure Sockets Layer Salute protocol.

Correct answer = B

SSL uses a combination of the SSL Record protocol and the SSL Handshake protocol to provide security. The Handshake protocol provides authentication services while the Record protocol provides for a secure connection. 802.1X uses the Extensible Authentication Protocol (EAP), which enables the technology to work with wireless, Ethernet, and Token Ring networks. Extensible Handshake Sockets Protocol and Secure Sockets Layer Salute Protocol are invalid selections.

7. **Please select the recommended ways to help secure your home or office network from Instant Messaging vulnerabilities? (Choose three)**

  ☐ A. Purchase and install an intra company messaging system.
  ☐ B. Disallow at-home VPN connections.
  ☐ C. Disable IM client FTP capabilities.
  ☐ D. Scan IM file transfers with your corporate antivirus solution.
  ☐ E. Install a personal firewall at every desktop.

Correct answers = A, C, and E

A, C, and E are in fact, correct. Although disallowing VPN connections would reduce the threat of infiltration of a corporate network from a home system, it is not a very productive solution and not recommended. An antivirus solution does not offer the ability to scan IM file transfers.

8. What is the proper UNC path to access a share named security on a server named CompTIA?

○ A. \\security\CompTIA.

○ B. \CompTIA\\security.

○ C. \\CompTIA\security.

○ D. \security\\CompTIA.

Correct answer = C

The correct answer is \\CompTIA\security. All other choices are invalid.

9. What is the 8.3 associated file name for the long file name Security-plus.DOC?

○ A. *Securi~1.DOC.*

○ B. *Security+.DOC.*

○ C. *Secur~1.DOC.*

○ D. *Securit.txt.*

Correct answer = A

For backward 8.3 file name compatibility newer operating systems automatically remove any space or invalid character and truncate the filename. Please refer to the Long File Names (LFNs) section in the chapter for more examples. All other choices are invalid.

10. Using 802.1X authentication, what names are used to identify the client and the access point?

○ A. Port requestor and port enabler.

○ B. Requestor and acceptor.

○ C. Applicant and authentication server.

○ D. Supplicant and authenticator.

Correct answer = D

With 802.1X authentication, a wireless client that wishes to connect to and be authenticated on a network is called a supplicant. The supplicant must first request access from an access point, which is also known as an authenticator. All other choices are invalid. The authentication server in 802.1X is not called the authenticator.

11. **When you access a Web page on the Internet, what can be placed in your system that is used as a sort of tracking device to uniquely identify you?**

    ○ A. Acceptor.
    ○ B. Cookie.
    ○ C. A Multipurpose Internet Extension.
    ○ D. UNC.

    Correct answer = B

    A cookie is a unique identifier that Web servers place in your system to identify you. Choices A and C are invalid. A UNC name is used to access a particular share on a particular workstation or server on a network.

12. **Which of the following are potentially harmful to the welfare of a system?**

    ○ A. Cookies.
    ○ B. Signed applets.
    ○ C. Java scripts.
    ○ D. All of the above.
    ○ F. None of the above.

    Correct answer = D

    Many Web servers use trusted cookies as their only form of authentication. This is a major security risk. Hijackers and attackers often create scripts and applets, which at times are able to circumvent network security parameters. They can be used to manipulate files on users' computers. Choice F is invalid.

13. **Concerning applet security levels, please select the choice that best represents the most restrictive to the least restrictive applet security level.**

    ○ A. Low, Untrusted, Medium, High.
    ○ B. Untrusted, High, Medium, Low.
    ○ C. High, Medium, Low, Untrusted.
    ○ D. None of the above.

    Correct answer = B

Applets can be assigned various security levels, which include Untrusted, High, Medium, and Low levels of security. Please refer to Table 3.1 for the various levels of unsigned applet security and their levels of restriction. All other choices are invalid.

14. **There are known vulnerabilities that exist with certain versions of LDAP. Name three results of these vulnerabilities mentioned in this chapter.**

    ○ A. Man-in-the-Middle, Backdoor7, Netbus.Patcher.

    ○ B. Backdoor7, Netbus.Patcher, Netbus.2.Trojan.

    ○ C. Buffer overflow attacks, unauthorized access, denial of service.

    ○ D. None of the above.

    Correct answer = C

    Known vulnerabilities exist with certain versions of LDAP that have lead to buffer overflow attacks, unauthorized access conditions, and Denial of Service. You will most likely face questions very similar to this on the real exam. Again, do not read too far into the question. It is very possible that you might see the other choices in this question on the real exam. Know your viruses for the exam!

15. **What is typically used to resolve a domain name to an IP address?**

    ○ A. Forward DNS.

    ○ B. DNS.

    ○ C. NDS.

    ○ D. None of the above.

    Correct answer = B

    DNS (Domain Name Server or Service) is used to resolve fully qualified domain names to node or IP address. Choice A is invalid. NDS (Novell Directory Service) is a Novell product used for managing users and resources on a Novell network.

16. **Excessively Long file names are known to cause what?**

    ○ A. Errors on UDP port 69.

    ○ B. Eight dot three errors.

    ○ C. Man-in-the-Middle attacks.

    ○ D. Buffer overruns.

    Correct answer = D

Exceptionally long file names are often intentionally used to exploit network software vulnerabilities and cause buffer overruns. All other choices are invalid.

17. **What is WEP used for?**

   ○ A. Wireless security.

   ○ B. Protocol used for connecting to NT 4.0 Server.

   ○ C. Used to connect a VLAN to a wireless network.

   ○ D. Security replacement for a secure access point.

Correct answer = A

WEP is a wireless security protocol intended to provide a WLAN with a similar security level as the protection that can be found in traditional LANs. RAS is a service provided with Windows NT 4.0, 2000, and XP that allows remote clients to access services and resources located on a network using an analog modem, WAN, or ISDN connection. Answers C and D are invalid.

REFERENCES

http://www.microsoft.com/windowsxp/pro/techinfo/planning/wirelesslan/solutions.asp is a Microsoft Web site that describes Wireless LAN security and 802.1X.

http://www.vmyths.com is a Web site that provides a wealth of information regarding computer virus myths and hoaxes.

http://wp.netscape.com/newsref/std/cookie_spec.html is a Netscape Web site that provides an excellent overview of cookies.

http://support.microsoft.com/default.aspx?scid=KB;EN-US;Q154036 is a Microsoft Web site describes how to disable active content in various versions of Internet Explorer.

http://www.java.sun.com/sfaq/#prevent is a Sun Microsystems Web site that answers many questions regarding Java and applet security.

http://www.w3.org/Security/Faq is an awesome site hosted by the World Wide Web Consortium. It provides a wealth of information regarding Web security.

http://www.orthus.com/ttvuln.html is hosted by Orthus Information Security Solutions. It provides an informative list of the top-ten most exploited Internet security flaws.

CHAPTER 4

INFRASTRUCTURE SECURITY

IN THIS CHAPTER

- IEEE 802 Specifications
- Open Systems Interconnection (OSI) Model and Layers
- Network Topology
- Network Cabling
- Removable Media
- Protocols and Services
- Intrusion Detection Systems (IDS)
- Network Countermeasures
- Devices
- Security Baselines
- Test Tips
- Chapter Summary
- Review Questions
- References

It is very important for anyone studying network security to be familiar with basic networking concepts and methodology. It is extremely important that you have a working knowledge of the material discussed in this chapter before you take the Security+ exam. We will begin our study of communications and networking with the industry standards and methodologies by which most networks are engineered. Then, we will focus on the physical layouts, components, and protocols used in most networks today. Finally, we will focus our attention on protecting networks from internal and external harm. If you are already familiar with some or all of the topics in this chapter from past experience, consider this chapter an important review of your networking knowledge. It is likely that the exam will drill you with questions from many of the topics discussed in this chapter.

IEEE 802 SPECIFICATIONS

The Institute of Electrical and Electronic Engineers (IEEE) is a technical organization or society that develops standards for local area networks (LANs) and wide area networks (WANs). The IEEE 802 project standards were developed in the 1970s as a set of specifications or rules that manufacturers and users can use as a sort of road map for understanding and developing networks and network related devices. The IEEE specifications are associated with certain networking layers of the OSI (Open Systems Interconnection) networking model, which will be discussed shortly in this chapter.

Project 802 was developed to address standards for network interface cards, network cables, and wide area networks. Many additions have been made to the IEEE 802 standards as technology has progressed.

The original 12 categories of the 802 specifications and their associations are as follows:

- **802.1:** Internetworking.
- **802.2:** Associated with the Logical Link Control (LLC) sublayer of the OSI networking model.
- **802.3:** Carrier-Sense Multiple Access with Collision Detection (CSMA/CD) local area networks (Ethernet).
- **802.4:** Token Bus LAN.
- **802.5:** Token Ring LAN.
- **802.6:** Metropolitan Area Network (MAN).
- **802.7:** Broadband Technical Advisory Group.
- **802.8:** Fiber Optic Technical Advisory Group
- **802.9:** Integrated Voice and Data Networks.

- **802.10:** Network Security Technical Advisory Group. Key management.
- **802.11:** Wireless Networking.
- **802.11a:** Applies to 5 GHz Wireless Technology.
- **802.11b:** Applies to 2.4 GHz Wireless Technology
- **802.12:** Demand Priority Access Lan100Base VG-AnyLAN.

Later in this chapter, we will discuss topologies such as bus, star, and ring. It is important to remember that the 802.3 Ethernet standards apply to bus and star networks that utilize CSMA/CD (Carrier-Sense Multiple Access with Collision Detection) access methods, while Token Ring topologies utilize token-passing methods to place a data signal on a wire.

The IEEE 802 standards apply mostly to the physical aspects of networking components. For example, they have to do with how a network interface card (NIC) is connected to a network and the particular types of media transmission methods used to carry a signal down a physical wire.

Just about every networking component manufactured today is designed to meet one of the earlier mentioned IEEE standards. If you purchase a hub, cable/DSL router, NIC, or wireless network component, take a look at the specifications on the packaging or in the advertisement. You will see that the product was manufactured to meet one of the standards set forth by the IEEE. If you are interested in learning more about the IEEE and its standards, you can visit the IEEE at *http://standards.ieee.org/*.

It is not likely that the Security+ exam will focus on the IEEE 802 standards. However, you should understand their underlying concepts and how they relate to the Open Systems Interconnection (OSI) model discussed next.

OPEN SYSTEMS INTERCONNECTION (OSI) MODEL AND LAYERS

As networking of computer systems became more popular in the world, a structured well organized logical framework for connecting networks and developing applications was needed. In the late 1970s, the ISO (International Standards Organization) developed the OSI networking reference model.

The OSI reference model is a seven-layered logical approach to network communications that includes specifications for the actual hardware connection to the network at the bottom layers of the model, and rules for applications and more complicated functions at the higher layers.

Networking rules for communication, also known as *protocols*, exist at almost every layer of the OSI model. The more complicated the protocol, the higher up on the model it resides. Important security-related protocols will be discussed in detail later in this chapter. Network transmission, security, session connection information, and hardware are all associated with particular layers.

Picture yourself sitting at your computer working on a Microsoft Word document. You are actually utilizing functions that reside at the top layer of the OSI reference model known as the Application layer or layer 7. You decide to attach your Word document to an e-mail message and send it to a coworker. The message or signal that you are sending is directed from the Application layer (layer 7) down through the other layers to the Physical layer (layer 1) where it is placed in converted format (0's and 1's) onto a network medium, such as a wire, and transmitted to your coworker. Your coworker, on the receiving end, accepts the message through their Physical layer (layer 1). The message is converted back to a readable format from 0's and 1's and is presented to your coworker's Application layer (layer 7).

The seven layers of the OSI reference model are as follows:

- **7 (Application layer):** *Application layer* applications, e-mail, FTP, user authentication, and any other major services that the end user interacts with directly are associated at this high-level layer. Network access and forms of error recovery are handled at this layer. High-level devices such as gateways are present at this layer. Application-specific protocols such as X.500, SMTP, SNMP, Telnet, and SMB reside at this layer as well as at the Presentation and Session layers (6 and 7, respectively).

- **6 (Presentation layer):** Data on the sending computer is converted into a format that can be transmitted over media to another computer. On the receiving end, data is converted into a format that the end user or Application layer can understand. Encryption and data translation occur at this level. The network redirector operates at this level.

- **5 (Session layer):** This layer establishes, holds, and controls sessions or connections between two applications. It provides checkpoint and synchronization service between two communication sessions. Security between two sessions is also handled at this layer.

- **4 (Transport layer):** This layer's primary concern is flow control and data handling. Large forms of data are broken down into manageable packets that can be presented to the higher layers on the receiving end. The successful transmission of data is acknowledged at this layer. If the transfer of information is incomplete or interrupted, this layer is responsible for a request of the information to be retransmitted by the sending application or session. Transfer protocols such as TCP, NetBEUI, NWLink, and SPX reside at this layer.

- **3 (Network layer):** This layer is responsible for the routing of information to the correct network, device, or computer. Logical names are converted to physical names at this layer. In other words, computer IP (Internet Protocol) addresses are converted to their MAC (Media Access Control) equivalents. Priority of connection and quality of service are also handled at this layer. A network router and switch reside at this layer. Network protocols such as IP and IPX reside at this layer.

- **2 (Data Link layer):** Data frames received from the network layer are converted into bits (0's and 1's) in preparation for the Physical layer (layer 1). On the receiving end, bits are packaged together into frames that can be understood by the higher layers. Frame synchronization, flow control, and error handling are addressed at this level. The Data Link layer has two sublayers known as the LLC (Logical Link Control) and MAC (Media Access Control) layers. The LLC sublayer is associated with IEEE standards 802.1 and 802.2 and is responsible for the implementation and placement of SAPs (service access points). The MAC sublayer is associated with IEEE standards 802.3, 802.4, 802.5, and 802.12. The MAC sublayer communicates directly with a network interface card. It is responsible for error-free communication between network interfaces. Devices called *bridges* segment network traffic and operate at this layer.

- **1 (Physical layer):** This is the physical adapter or connection to the network wire or medium. This is where bits or bit streams of information are prepared to go across the network medium. Incoming bits of information are organized and prepared to move through the higher layers. Networking devices such as repeaters and hubs operate at this level.

If you are interested in a career in networking, it is very important that you understand the theory behind the OSI reference model. You will also need a solid understanding of the OSI reference model and the IEEE specifications if you wish to understand how networking components, protocols, and applications function.

Although the OSI reference model is not listed in the CompTIA Security+ objectives, you'd better be prepared to identify the various layers at which particular network devices and protocols operate.

NETWORK TOPOLOGY

A network's *topology* is the actual physical layout of the network. The topology of a network is based on factors such as the number of workstations and servers required, the communication methods that will be implemented, and the cables and specialized equipment that are available. As you will learn soon, network-related security issues also play a major role in the way networks are constructed physically.

The three standard network topologies are as follows:

- Bus
- Star
- Ring

By the end of this chapter, you should be able to identify the three main typologies and the cable types associated with each.

The three topologies and their associated characteristics form the framework from which most networks are based. All three topologies are described in the following sections.

BUS

An Ethernet *bus topology*, otherwise known as a *linear bus*, is a topology designed for a limited number of computers that are typically attached to a single wire or *trunk* in a straight line. As you add more workstations to a bus topology, performance decreases. Figure 4.1 displays a typical bus topology.

The main type of cable implemented in a bus topology network is 10Base2. Devices called *terminators* must be placed on both ends of a bus network or wire to keep the data signal placed on the wire from bouncing back and forth. This phenomenon is referred to *signal bounce*. All computers on the bus listen for the data signals. If the signal is addressed to a particular workstation, the workstation accepts the signal. If the physical wire or connection that makes up the bus topology is damaged or breaks, the individual computers on the bus will still be able to operate independently

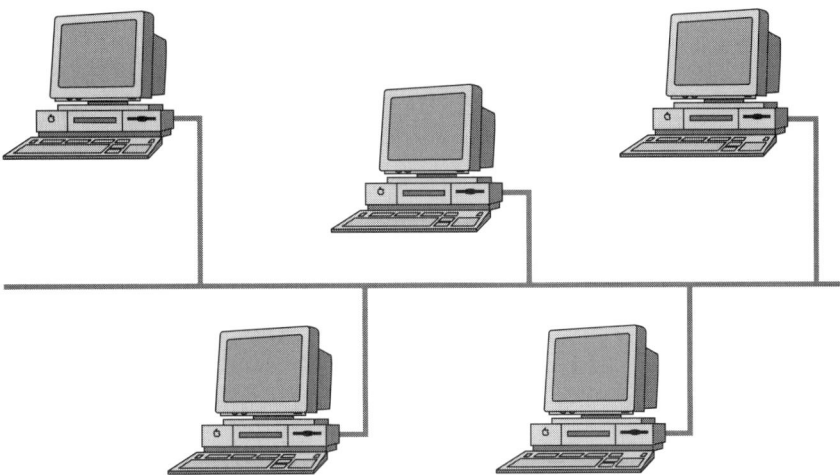

FIGURE 4.1   A linear bus topology network.

but will not be able to accept data signals and communicate with other computers on the bus network.

Bus and star networks utilizes CSMA/CD media access control methods to place a signal on a wire. CSMA/CD is an Ethernet error detection method used to ensure proper data handling on a wire.

BNC and BNC barrel connectors are used to attach a bus cable to a device or connect one piece of the bus cable to another. A device called a *repeater* is used to boost or regenerate the signals placed on a 10base2 or bus network. Adding a repeater to your bus can extend the length of your entire bus network. For distances associated with a bus network, refer to the cabling section later in the chapter.

The following points are important to remember regarding a linear bus topology:

- Use less total cable length than a star topology network.
- Failure of one computer fails does not prevent the network from functioning. If there is a break or failure on the main bus cable, the entire network will fail.
- Terminators are required at both ends of the bus backbone cable.
- Problems are difficult to isolate on a bus network if the entire network fails.

STAR

A *star topology* physically looks like a star. See Figure 4.2 for a simple star topology network. A star topology utilizes a central device, which can be a hub, a router, or a switch. These components will be discussed in detail later in the chapter. For now, we will refer to a simple hub for the central connection point in a star network.

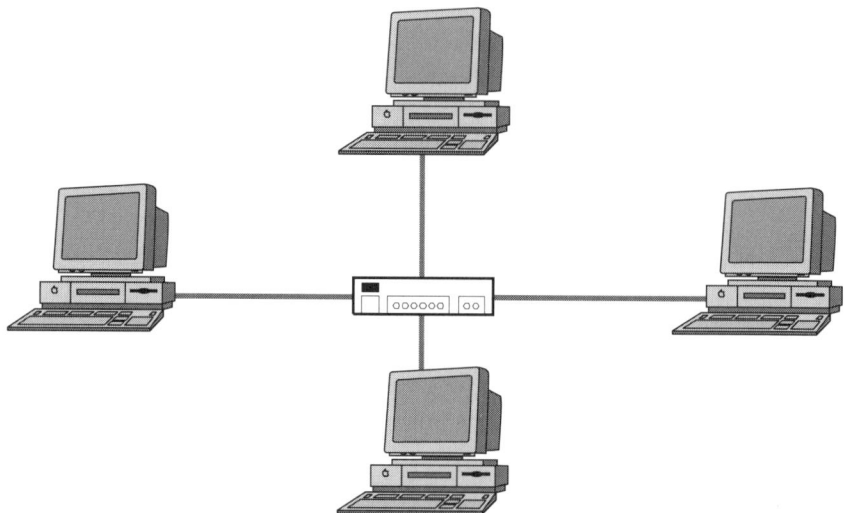

FIGURE 4.2   A star topology network.

All devices in a star topology network are typically connected to a hub with twisted pair, otherwise known as 10BaseT cable. A star topology network is known to require large amounts of cable for larger networks. The hub provides a central location where the network can be managed and tested. If one computer fails on a star network, the other computers connected to the hub can still function and communicate with one another. However, if the hub fails, all communication between devices will cease.

The following points are important to remember for the exam regarding a star topology network:

- Use a centralized hub, router, or switch to manage network traffic.
- Failure of the central connection device causes the entire network to fail.

- Failures are easier to troubleshoot. It is also easier to add and remove devices from the network through use of a central connection point.
- Expense is greater than a linear bus network.

RING

A *ring topology* network is best understood by picturing an actual circle of cable. Workstations and servers are all connected to the circle of cable. There is no end to the ring network that must be terminated. Each computer attached to the circle regenerates the data signal sent in the form of a token. The circling of the token to each of the servers and workstations on the ring is known as token *passing*. If the wire that makes up the ring is damaged, all network activity on the ring will cease. Keep in mind, IEEE 802.5 is a standard that applies to token passing technology. Figure 4.3 displays a simple token ring topology.

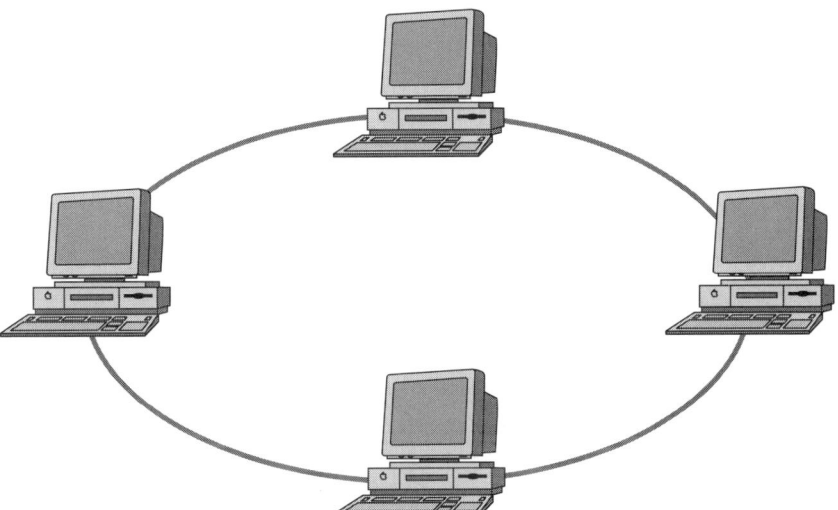

FIGURE 4.3   A ring topology network.

The following points are important to remember regarding a ring topology network:

- Nodes (systems) attached to the ring act as repeaters that boost the signal's strength as it travels around the ring.

- Breaks in the ring's structure cause the entire network to fail. If a network interface card (NIC) connected to the ring goes into an error state or is set to the wrong ring speed, the entire ring might fail.
- Ring technology is considered expensive.

FDDI (FIBER DISTRIBUTED DATA INTERFACE) RING

An *FDDI (Fiber Distributed Data Interface)* ring is based on a set of ANSI (American National Standards Institute) standards for data distribution over fiber optic cable in a network. An FDDI ring is typically composed of two fiber optic token rings. An outside ring is used for the primary transport of data and an inside ring acts as a backup if the primary ring fails. The following are important facts to remember regarding FDDI rings:

- Serve often as network backbones for WANs.
- Can support thousands of users.
- Capable of up to 100Mbps data transmission speeds.
- Travel of data in opposite directions in a dual FDDI ring topology— one ring clockwise, the other ring counterclockwise.
- Use of token passing technology.
- Has a distance capability of up to 200km (124 miles) in a LAN environment.

MESH

In a *mesh topology* network, all nodes (systems) are connected to each and every other node on the network (Figure 4.4). If the connection from one node to another node fails, a different route can be taken to access other nodes on the mesh network. This redundancy makes a mesh network one of the most reliable topologies available. Mesh networks are classified into two types of topologies. The more expensive but more redundant full mesh topology and the less expensive not as redundant partial mesh topology. Typically, a partial mesh topology is used as a backbone to connect full mesh topology networks.

INTRANET

An *intranet* is a network that is considered private and separate from the outside world. It exists to connect the workings of an internal network. Typically, separate internal networks or LANs are connected to a larger in-

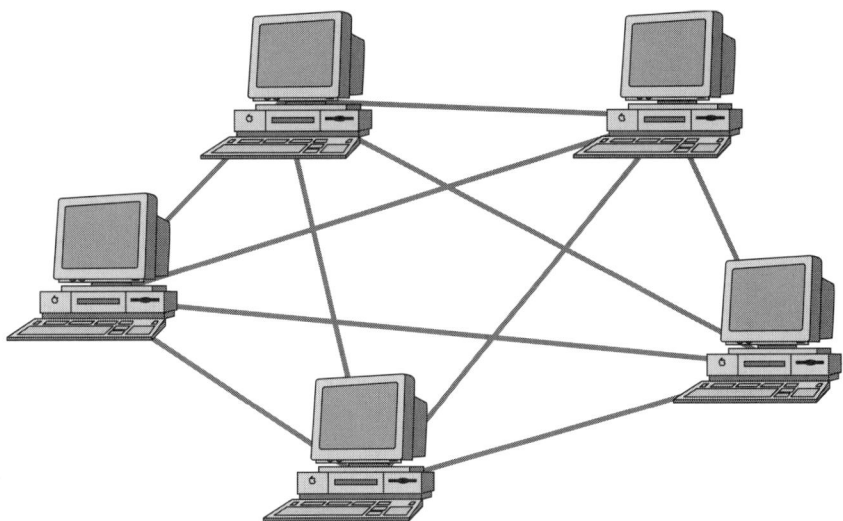

FIGURE 4.4   A mesh topology network.

ternal WAN through dedicated leased lines using connections such as T1 or DS3. Gateways and firewalls are usually implemented to allow users within an intranet to connect outside of the network.

EXTRANET

Allowing outside vendors, suppliers, and clients access to internal network resources and information has become necessary to and productive for the daily operations of most modern businesses. When part of an internal network or intranet has been made accessible to outside sources, that part of the internal network is referred to as an *extranet*. Access to the internal network typically requires use of public telecom technologies as well as a high level of security. This usually includes encryption and digital certificate technologies, as well as the use of firewalls.

VLAN (VIRTUAL LAN)

A *VLAN* is a network of systems where any single system can reside on a separate network but be configured as though it were on the same network as other systems in the VLAN. In other words, systems in a VLAN appear and work as though they are all on the same subnetwork. In reality, they are located on separate or different subnetworks. A great advantage to using a VLAN is that systems can be moved within a VLAN without having to make hardware adjustments. Companies such as 3com and Cisco Systems offer VLAN software.

NETWORK CABLING

There are over 2,500 types of cable in use to connect computers and peripherals. The majority of computers in use today still use some type of wire or cable to transmit data from one system to another. There are three primary types of network cables in use that you need to be familiar with:

- Coaxial
- Twisted pair
- Fiber optic

Each of these cable types has different characteristics that set them apart from the others such as cost, distance limitations, data transfer methods, data transfer rates, and installation methods used.

COAXIAL CABLE

Coaxial cable is separated into two distinct categories: thicknet and thinnet.

Thicknet

Thicknet coaxial cable, otherwise known as *10Base5*, is approximately 0.5 inches thick; it is a heavy type of cable with a copper core that was used with early mainframe computers and early networks. Thicknet coaxial still exists but is very limited in its ability to achieve high data transfer rates that are needed to support today's bandwidth-hungry computers and applications.

Thicknet coax cable has the ability to carry 10 megabits of data a total distance of 500 meters or approximately 1,500 feet. Thus, the naming convention scheme of 10Base5 has been established for coax cable. In other words, 10 megabits of information can travel over a baseband medium or base a total of 500 meters; 5 multiplied by 100 equals 1,500 feet but the true measurement is closer to 1,640 feet. The naming convention drops the last two 0's.

Thicknet coaxial cable was and sometimes still is used as a backbone connection that connects to a small thinnet cable by use of a Vampire Tap and an AUI (Attachment Unit Interface) connector.

Thinnet

Thinnet coaxial cable, otherwise known as *10Base2*, is approximately 0.25 inches thick. It is a thinner, more flexible type of coaxial cable that is usually connected directly to a NIC with a BNC or BNC T-connector. Thinnet is much easier to install and work with than thicknet. However, thinnet only carries a data signal the distance of 185 meters or approximately 607 feet.

Both thicknet and thinnet coaxial can make up a network referred to as a bus network. A bus network must be terminated at both ends of a cable or the bus network will fail. Thus, thicknet and thinnet both require terminators at both ends of their cable.

TWISTED-PAIR CABLE

Twisted-pair cable, or *TP* for short, arose from the need to replace the distance and other limitations associated with coaxial type cable. TP is referred to as *10BaseT*. Once again, the *10* is for the transmission rate of data, the *base* refers to a baseband media type, and the *T* refers to the twisted pair or wiring twists in the cable itself.

There are two types of twisted pair wiring: shielded twisted pair (STP) and unshielded twisted pair (UTP).

Shielded Twisted Pair (STP)

Shielded twisted pair is basically the same type of wire as unshielded twisted pair with the exception that STP uses a woven copper braided shielding and foil wrapping that protect the twisted wire pairs from outside interference such as EMI (Electro Magnetic Interference). This also allows an STP wire to be less susceptible to crosstalk from other wires. STP is more expensive than unshielded twisted pair based on its extra protection and ability to transmit a data signal over a longer distance than UTP.

Unshielded Twisted Pair (UTP)

Unshielded twisted pair, or UTP for short, is also a 10Mbps baseband cable. UTP, generally referred to as *10BaseT*, is the most common type of Ethernet cable in use today and is found mostly in star typology networks. UTP in its simplest form is two insolated copper wires that can carry a data signal 100 meters or approximately 328 feet.

There are five categories for UTP wiring as specified by the Electronics Industries Association and the Telecommunications Industries Association (EIA/TIA) to keep wiring standards uniform. They are as follows:

- **Category 1 (CAT1):** This is the original implementation of UTP used for telephone cable. It is capable of transmitting voice but not data. This type of phone wire was installed prior to the mid-1980s.
- **Category 2 (CAT2):** This UTP cable type is made up of four twisted pairs of wires. It is cable of transmission rates up to 4Mbps.
- **Category 3 (CAT3):** This UTP cable can transmit data up to 10Mbps. It has four twisted pairs that are twisted three times per foot.
- **Category 4 (CAT4):** This UTP cable is capable of data transmissions up to 16Mbps. It has four twisted pairs of wire.
- **Category 5 (CAT5):** This UTP cable is capable of data transmission rates of up to 100Mbps. It is also made of four twisted pairs of wire. CAT5 UTP is also referred to *100BaseT* or *100BaseTX*. It carries a data signal 100 meters or approximately 328 feet. It is the most popular UTP cable in use today. CAT6 and CAT7 UTP are also available today and offer transmission speeds up to 155MBps and 1GBps, respectively.

TWISTED PAIR CONNECTORS

There are two types of UTP connectors you should be familiar with: RJ-11 and RJ-45. An *RJ-11* phone connector is used for early categories of UTP to connect a modem to a typical phone jack or your phone to a phone jack. In technical circles, an RJ-11 wire is a simple phone wire that houses four wires or connections.

An *RJ-45* connector is the most common type of TP data cable connector in use. It houses eight wire traces. The RJ-45 connector on one end of a TP wire plugs into a network interface card that is installed into a system. The RJ-45 connector on the other end of the TP cable plugs into a network hub, router, or RJ-45 wall jack.

CROSSOVER CABLE

A *crossover cable* is a type of Ethernet TP cable that is commonly used to connect computers in a peer-to-peer fashion. The crossover cable switches the transmit and the receive lines of the cable, which allows two computers to communicate directly with each other without the use of a hub or router. If you want an inexpensive alternative to purchasing a hub, a crossover cable is the way to go to connect two computers.

FIBER-OPTIC CABLE

Fiber-optic cable, otherwise known as *10BaseFL,* is the network wire of choice. It is capable of extremely fast transmission rates over long distances without interference. A fiber-optic cable has a core that is composed of plastic or glass. A glass cladding or sheath covers the core. Finally, a Kevlar fiber jacket surrounds the entire wire. Data can be transmitted through a fiber-optic cable with a laser or LED (Light Emitting Diode) at a rate of 2Gbps or higher. The data signal on a fiber wire can travel up to a distance of 100 kilometers or 60 miles depending on which technology is being implemented with the fiber and if a repeater is used.

Fiber optic cables use special ST and SC type connectors to attach to network interface cards and fiber optic ports. These connectors are precisely crafted and specially designed to suit fiber-optic cable connection requirements.

Fiber-optic cable needs great care and consideration when being installed. Specially trained and certified fiber installers are usually empowered to carry out this task. Because of its high transmission speeds and specialized installation methods, fiber-optic technology is quite expensive.

Fiber-optic cable is much more difficult to tap into than other types of network cable. In other words, it is considered the most secure type of network cable. Special equipment and skilled hands are required to carry out such a task. This is not the case with twisted pair and coaxial types of cable.

Refer to Table 4.1 for a comparison chart of the major networking cables described in this chapter.

TABLE 4.1 Cable Comparison Chart

Cable Type	Transmission Speed	Distance
10BaseT	10Mbps	100M/328ft
10Base2	10Mbps	185M/607ft
10Base5	10Mbps	500M/1,500ft
100BaseT	100Mbps	100M/328ft
Fiber	100Mbps to 2Gbps	100K/60 miles

REMOVABLE MEDIA

There are many different types of removable media available today. Most types of removable media are used for storage, memory, or network connectivity purposes. It is likely that the Security+ exam will expect you to have a basic understanding of the following media types as well as knowledge of the proper procedures for backing up and securing each of them. It is most important that you remember there is no replacement other than a good backup system to ensure that your data will always be available to you. New viruses and mechanisms are constantly being developed to attack your system and steal or destroy your programs and files. Always back up your important media.

TAPE

Tape is magnetic media on which data can be stored or read. In combination with a tape unit and software, tape can be used to back up information from stand-alone or networked computer systems. Accessing data stored on tape media is considerably slower than using other storage media, such as a hard drive. However, because of their portability, tapes offer a good mobile solution for storing data off site in the event of an emergency. (More details regarding backups and off-site storage will be discussed in Chapter 6). Unfortunately, because of their portability, tapes and their valuable data can also be easily stolen. You should always secure your tapes in a locked box or cabinet located in a secured area.

In regards to securing the data stored on your tapes, any good software package used for backing up data to tape usually includes an encryption scheme that you can use to protect your stored data. Keep in mind, most tape backup encryption methods will affect the performance of your tape backup while it is in the process of backing up data.

There are several types of tape available today. Some of the most popular are as follows:

- **DLT (digital linear tape):** Used to store large amounts of data at very high data transfer rates.
- **DAT (digital audio tape):** Used as a standard medium for recording audio. DAT and DAT units provide a simple and easy-to-use archival system. They are portable and compact.

Tape backup methods, rotation, and disaster recovery methods will be discussed in Chapter 6.

CD-R

A *CD-ROM (CD-R)* is an optical storage disk capable of storing large amounts of data. A typical CD-R can hold 600 to 800MB of information. This is equal to the storage capacity of about 700 1.44mb floppy disks. CD-Rs are well suited for storing graphics files, movies, and music. A CD-R is typically written to or burned once with information provided by the manufacturers of the CD-R. Optical media such as a CDs have their information burned into them by a laser beam. Actually, the term *burned* is used quite loosely. The actual process that takes place for information to be written to a CD involves the changing of the reflective properties of an organic dye that covers a CD by use of a laser. The data can be written to only one disk one time. Reading the data on a CD-R requires the use of a CD-R device or player.

If you are concerned about securing the data stored on CD media, your best bet is to use a third-party utility to encrypt the data that you plan on storing on the CD. If you use CDs in your backup scheme, you should include them in your off-site storage backup rotation plan.

 You should know the various types of removable media and the security risks involved with them. Fair warning: the exam is likely to target methods of media and data backup. These might include incremental, differential, and full backups. All of these are detailed in Chapter 6.

HARD DRIVES

Removable hard drives offer the flexibility of making large amounts of data mobile. They are also an excellent alternative to partitioning if you wish to run multiple operating systems from a single system unit. Many businesses and schools implement the use of *hot swappable* removable hard drives for training programs. You can work on your operating systems and applications, then simply remove the disk from the system and take it with you. The obvious problem here is that the flexibility and mobility of the drive make it very easy to steal. Most removable drives offer a lock and key, which can be used to secure the drive into the system unit. Unfortunately, many of these locks can be easily picked and most of the keys are interchangeable.

File and Folder Encryption

The most popular method of securing data on hard drives is through the use *encryption*. Most new operating systems in use today offer the ability to encrypt data and there are many third-party hard drive security encryption tools available that can help you protect your sensitive data.

Windows 2000 offers the *EFS (Encrypted File System)*, which allows you to encrypt files and folders located on your hard drive. EFS has many benefits. It is very secure and particularly useful for securing laptops and systems that are exposed to the general public. It is very easy to implement, allowing even novice users the ability to use public key encryption. It also offers the ability to have files automatically encrypted when they are stored within a folder. It is important to keep in mind that you can implement EFS only on a drive or partition that has been formatted with NTFS (NT File System.)

DISKETTES

A *diskette* or *floppy disk* is a magnetic portable storage disk that is often used to store small amounts of data. The major advantage of using a floppy disk is mainly that it is portable. You can move or copy a file or several small files to a floppy and take them with you.

There are two basic forms of floppy disk media available:

- **5.25-inch:** This older style floppy disk was popular in the 1980s. The 5.25-inch floppy came in two common sizes with storage capabilities of 360K and 1.2MB of data. The 5.25-inch floppy disk used a 5.25-inch floppy drive that is now considered obsolete.
- **3.5-inch:** This floppy drive is very common in most computers today. The 3.5-inch floppy disk can store 720K (double density) or 1.44MB (high density) of data.

Because floppy disks or diskettes are portable, they provide a security risk. An attacker can physically insert a floppy disk into your system, copy data to the floppy, and walk away. One way to combat this issue is to have floppy-less systems installed in your "open-to-the-general-public" production environment, or use an operating systems policy that hides or disables use of the physical floppy drive.

If you need to use your floppy drive and floppy diskettes, you should always write-protect your floppy disk after you have written data to it and removed it from your floppy drive. You should also make a backup copy of your important floppy disks. There is no substitution for a good backup. If you lose your data on your original disk or your original media malfunctions, you can always restore from your backup disk. As a final note, you can protect the files on your diskettes with simple encryption techniques to ensure that your data is secure.

FLASH CARDS

A *flash card* or *memory card* is a small portable storage card that has *non-volatile* memory, meaning that it will not lose its data when disconnected from a power source. Flash cards are most often used with portable computing devices such as digital cameras, PDAs (Personal Data Assistants), and game controllers. They are commonly used to store data such as text, graphics, audio, and game files. Because they are small in size and can be easily concealed, the obvious security risk with flash cards is that they can be easily stolen.

SMART CARDS

A *smart card* (another form of memory card) is a plastic card or token device approximately the size of a credit card that houses a microchip. A smart card can be used in the following ways, in addition to other functions that are currently under development:

- Data storage and memory purposes.
- Security authentication device to allow access.
- Telephone-dialing access.
- Electronic payments.

Smart cards can also contain important personal information such as bank and credit card information (PINs), biometric details, cryptic key information, and other sensitive data.

A smart card is typically inserted into a smart card reader that reads the data stored on the card. Because of their self-contained design and built-in security features, smart cards are considered to be one of the most secure storage devices available today.

 Encrypted keys can be stored on smart cards. Be sure your smart card is stored in a secure place or you could be in a heap of trouble with a client and on the Security+ exam!

PROTOCOLS AND SERVICES

Network *protocols* are defined as a common language or set of rules that computers use to communicate with one another. Protocols come in packages known as *protocol stacks*. Individual protocols reside at each layer of the OSI reference model in order to carry out specified functions. In this section, we will discuss the most commonly used communication protocols such as TCP/IP, IPX/SPX, and ARP. Then, we will focus on protocols specifically designed for security purposes.

Security protocols are typically used to encrypt and decrypt data packets for safe transmission over a communications medium.

TCP/IP

TCP/IP (Transmission Control Protocol/Internet Protocol) is the most popular protocol in use today. It is the protocol of choice for the Internet. TCP/IP is commonly used with Ethernet, Token Ring, and Internet or dial-up network connections. Every computer on a TCP/IP network uses an IP address as a unique numeric identification. An *IP address* is a 32-bit numeric combination of four period-delimited octets, each of which can be a number from 0-255. IP address can be up to 12 digits long.

An example of an IP address is 209.15.176.206. This IP address is associated with the domain name address, which is provided by a Domain Name Server (DNS) of the publisher of this book, *charlesriver.com*. From a DOS prompt, I used the TCP/IP ping utility to test my connection to the Charles River Media Web site. Try it. Using Windows 2000, navigate to a DOS prompt. At the DOS prompt, type **Ping charlesriver.com**. You should receive an associated IP address of the Web site as well as four echo replies.

Another popular TCP/IP utility is the tracert command. The tracert will tell the route you are using to establish a connection with a destination computer. In other words, it will give you all of the TCP/IP addresses and domain names of the computers you are using to reach your final destination. Try the tracert command from a DOS prompt. At the DOS prompt, type **tracert charlesriver.com**. You will receive the IP addresses and domain names of the computers you are hitting to get to the Charles River Media Web site. The time it takes for your request to go from each of the destination IP addresses you see is measured in units of time called *hops*. A subnet mask is used to specify which particular network a TCP/IP address belongs to.

You can check the IP configuration of your computer using two popular commands. If you are using Windows 95 or 98, type **winipcfg** at a DOS prompt. If you are using Windows NT or 2000, type **ipconfig** or **ipconfig /all** at a DOS prompt. Your computer's IP address, subnet mask, and default gateway settings will be displayed.

If your computer is unable to communicate with other computers on the network, and all of the other computers are functioning correctly, you should first check your computer's IP address configuration settings. They might not be properly configured. If this is the case, your computer will only be able to access itself.

The *IP* portion of the TCP/IP protocol is actually responsible for delivering messages and data over networks. The TCP protocol's main function is to package packets back together after they arrive at a destination. It is very important to note the most common versions of IP. They are: IPv4 (Internet Protocol Version 4) and IPv6 (Internet Protocol Version 6.) IPv6 is the latest version of IP. It is packaged with most newer operating systems of the day and in most cases has replaced the earlier IPv4 based on its major improvements. IPv6 provides rules and support for the following types of addressing:

- **Unicast:** Transmits rules for sending messages from a single host to another single host.
- **Anycast:** Involves rules for transmission from a single host to the nearest possible host from a grouping of hosts.
- **Multicast:** Involves rules for transmission from a single host to an entire group of hosts.

The most important fact to remember regarding these two versions of IP is that IPv6 allows for IP addresses to be lengthened from the IPv4 limitation of 32 bits to 128 bits.

IPv6 also allows better authentication, privacy, and improved data delivery assurance.

TCP/IP Address Classes

TCP/IP address classes are divided into five distinct classes as defined by the *InterNIC*. The InterNIC is a consortium of businesses whose purpose is to manage certain services for Internet users and business. One of the primary services provided by the InterNIC is the management and assignment of TCP/IP address classes. The five classes of TCP/IP addresses provided by the InterNIC are as follows:

- **Class A:** Used for large networks. Supports up to 16 million host systems on each of 127 networks.
- **Class B:** Used for medium size networks. Supports up to 65,000 host systems on each of 16,000 networks.
- **Class C:** Used for small to mid-size networks. Supports up to 254 host systems on each of 2,000,000 networks.

- **Class D:** Used for multicast service only.
- **Class E:** Used for experimental purposes only.

Address classes A, B, and C each have their own set or block of reserved IP addresses that are specifically used for private internal networks. These IP addresses are not routable addresses and cannot be seen or accessed by default beyond the boundary of the local network on which they are utilized. The reserved addresses for Class A networks are 10.0.0.0 through 10.255.255.255. The reserved addresses for Class B networks are 172.16.0.0 through 172.31.255.255. The reserved addresses for Class C networks are 192.168.0.0 through 192.168.255.255.

The Security+ exam just might ask you a question such as the following:

Which of the following IP addresses is reserved for internal use and cannot be used on the Internet?

A. 172.14.42.5.

B. 172.42.42.5.

C. 172.31.42.5.

D. 172.52.42.5.

The correct answer to this question is C. 172.31.42.5. Notice that all of the other choices do not fit into the range of privately reserved Class B. In other words, the IP address would have to fall between 172.16.0.0 and 172.31.255.255 to be considered private.

IPX/SPX

IPX/SPX (Internetwork Packet Exchange/Sequenced Packet Exchange) is a protocol stack used in Novell networks that supports routing. There are several versions of Novell operating systems in use today. When connecting a system to a Novell network, it is often necessary to bind a specific frame type to your network interface card for connection to various Novell operating systems versions. Frame type specifications are beyond our study focus. Just remember that if you have trouble connecting to a Novell network, you should first verify that the proper frame type is bound to your NIC.

ADDRESS RESOLUTION PROTOCOL (ARP)

Address Resolution Protocol (ARP) is a protocol used to map an IP (Internet Protocol) address at the network layer of the OSI model to a physical hardware address at the MAC (Media access Control) sublayer.

Packets that arrive at a gateway (typically a router) on a network are resolved by an ARP cache or table and directed to the proper destination system or subnetwork based on this IP to MAC resolution technique.

Every network interface card (NIC) has a hard-coded hardware or MAC address programmed to it or burned to its PROM chip by the manufacturer. ARP resolves the computer systems IP address to this hardware address. On a typical Windows 2000 system, you can view your computer system's IP, MAC address, and local ARP cache by navigating to a command prompt and entering **arp –a**.

The following are the results of entering arp –a on a system:

```
Interface: 10.1.18.67 on Interface 0x1000003
Internet Address Physical Address Type
10.1.18.1 00-00-0c-07-ac-0d dynamic
10.1.18.77 00-d0-b7-4f-22-f2 dynamic
10.1.18.137 00-10-5a-01-d1-2a dynamic
10.1.18.211 00-d0-b7-4f-55-1b dynamic
```

SECURE SHELL (SSH)

Secure Shell (SSH) is a UNIX-based strong authentication method used to allow administrators to access and control remote systems securely. SSH is actually a suite of newer UNIX protocols and utilities including ssh, scp and slogin that replace the older UNIX rcp, rsh, and rlogin utilities. Unlike these older utilities, password encryption and digital certificates are used with SSH to ensure the entire communication channel is secure from the client to server and server to client. This makes it virtually impossible for attackers using spoofing and IP source redirecting techniques to interfere with a communication session. We will discuss SSH in more detail in Chapter 5.

HIGH LEVEL DATA LINK CONTROL (HDLC)

High Level Data Link Control (HDLC) is a transmission protocol that operates at the Data Link layer (layer 2) of the OSI model. HDLC and *SDLC (Synchronous Data Link Control)* were originally developed for IBM SNA network architecture.

Today, HDLC is most commonly used in X.25 frame relay packet switching networks that are used by the Internet. HDLC establishes a session for communication where one node is designated as the primary node and another node is designated the secondary node. After this designation takes place, the following communication modes can be implemented:

- **Normal unbalanced mode:** The secondary node responds only to the primary node's request.
- **Asynchronous mode:** The secondary node can begin the communication by sending a message.
- **Asynchronous balanced mode:** Both stations can send and receive messages by duplex transmission. Networks that implement X.25 packet switching most commonly use this mode.

SYNCHRONOUS DATA LINK CONTROL (SDLC)

Synchronous Data Link Control (SDLC) is the original IBM developed communications protocol from which HDLC is based. SDLC is based on a primary/secondary communications model where a secure connection is established between a mainframe (host) and a client. Multiple clients can be connected to a common point with SDLC. This connection technique is known as *multipoint* or *multidrop*. SDLC is an efficient protocol for private networks with dedicated lines of communication.

PASSWORD AUTHENTICATION PROTOCOL (PAP)

Password Authentication Protocol (PAP) is a basic type of authentication where a username and password are transmitted unencrypted across a network to an authenticating host. The host houses a security table or database that is typically encrypted. If the requesting username and password match those stored in the server's database, an acknowledgement is sent to the requester and authentication is granted. In a Windows client/server environment, the server's security database is called the *SAM (Security Accounts Manager.)* PAP is an outdated authentication method. If CHAP (described next) is available on a server, it should be used in place of PAP.

 PAP is outdated. For better security, use CHAP.

CHALLENGE-HANDSHAKE AUTHENTICATION PROTOCOL (CHAP)

Challenge-Handshake Authentication Protocol (CHAP) is a much more secure method of authenticating communications between server or agent and requester than PAP. CHAP uses a secret one-way hash value that is generated by the requester and sent to the server. The sever builds upon the hash value and generates a secret *MD5* value that is only known to the requester and server. (MD5 is an algorithm used to create digital signatures.)

If the requester's value matches the server's hash value, the requester is authenticated. Only the hash value is transmitted during communication using CHAP.

 The CHAP authentication process is called a *three-way handshake*. It is highly probable that you will run across the CHAP authentication process several times on this exam.

POINT-TO-POINT PROTOCOL (PPP)

Point-To-Point Protocol (PPP) is a full-duplex serial communication protocol that operates at the Data Link layer of the OSI reference model. Internet service providers (ISPs) often use PPP to link their customers to the Internet. PPP has replaced the older SLIP (described next) based on its error-checking capabilities and ability to provide more stability.

SERIAL LINE INTERNET PROTOCOL (SLIP)

Serial Line Internet Protocol (SLIP) is an older TCP/IP communications protocol used to connect two computer systems. SLIP was and sometimes is still used to connect systems to the Internet through a slow analog dial-up connection. SLIP does not handle error detection as PPP does and is considered outdated technology.

X.25

X.25 is a communications protocol used in packet switching networks designed to handle the passage of analog data. The X.25 standard was adopted by the CCIT (International Consultative Committee for Telegraphy and Telephony) in 1976 as a solution needed to provide interconnection between different types of internal networks.

X.25 operates at the Physical, Data Link, and Network layers of the OSI reference model to provide reliable communication between such communication devices as a host or Data Terminal Equipment (DTE) and a node, known as *Data Circuit Terminating Equipment (DCE)*. X.25 is designed to support *Switched Virtual Circuits (SVCs)* and *Permanent Virtual Circuits (PVCs)*. SVCs are circuits that are set up between hosts and are only active as long as data is being transferred between the hosts. PVCs are permanent circuits that exist all of the time. Today, X.25 is used overseas more often than in the USA.

FRAME RELAY

Frame relay is a telecommunications service based on X.25 packet switching technology. Frame relay is much more effective than X.25 based on its ability to package data packets in variable sizes and leave error correction and other services up to service endpoints in communication. Frame relay is also much faster than X.25 and can take advantage of T1 (1.544 Mbps) and T3 (44.736 Mbps) speeds.

Frame relay technology is available through most phone service providers. It is considered inexpensive and is most commonly used to connect smaller LANs (local area networks) to larger WANs (wide area networks).

ISDN (INTEGRATED SERVICES DIGITAL NETWORK)

ISDN is a baseband transmission technology that is well suited for the transmission of audio and video at rates of up to 128Kbps. ISDN utilizes an adapter that is included with and ISDN router in place of a standard analog modem.

There are two types of ISDN services typically available by ISP or local phone carrier. They are as follows:

- **BRI (Basic Rate Interface):** This is an ISDN technology made up of two 64Kbps B-channels that carry data and voice, and a 16Kbps D-channel that is responsible for control information. BRI implementations are common for small business and home use.
- **PRI (Primary Rate Interface):** This ISDN technology is used with larger businesses such as ISPs and telecommunication companies. PRI is made up of 23 B-channels and one D-channel. PRI typically utilizes the bandwidth capabilities of a T1 connection.

DSL (DIGITAL SUBSCRIBER LINE)

DSL is a connection technology over a copper wire that utilizes a regular phone line to bring access speeds of up to 6.1Mbps to homes or businesses. In actuality, DSL offers upload speeds of up to 128Kbps and download speeds of 1.5Mbps for individual connections. DSL also utilizes a modem and is well suited for high-speed transmission of audio and video. DSL has provided major competition to the cable modem.

INTRUSION DETECTION SYSTEMS (IDS)

Intrusion detection systems (IDSes) are considered to be the first line of defense against attacks or odd behavior on networks and computer systems. They are like a centurion overlooking perimeter walls. When attacks take place, such as a hacker trying to gain unauthorized access to a network or a probe program searching for open network ports, the IDS or "centurion" typically alerts a network security guard or administrator that an attack has occurred. The most common methods of alert are by pager, SNMP trap messages, or e-mail.

Irregular activity in a network can also be detected by intrusion detection systems. A good IDS will work well for both external and internal threats, or intrusions and misuse of a network. First, we will discuss the two main categories or types of intrusion analysis— signature-based intrusion analysis and statistical-based intrusion analysis. Then, we will take a look at the two main types of intrusion detection systems known as *network-based IDS* or *NIDS* and *host-based IDS.*

SIGNATURE INTRUSION ANALYSIS

Most malicious attacks on networks and systems look for vulnerable areas to penetrate and eventually do their dirty work. Most attacks tend to develop patterns in the way they search for weak areas of a system or network. Developers of intrusion detection systems and software analyze the patterns and develop software and hardware IDSes that can detect these patterns and report that misuse or an intrusion has occurred.

Developers typically include code known as a *signature* into the IDS. In short, *signature analysis* is a method used to compare known attack patterns against a signature database that contains information about known attacks. If you are using a commercial product that implements this method of analysis, it is suggested that you keep the signature database up-to-date regularly with patches and updates from the manufacturer's Web site.

STATISTICAL INTRUSION ANALYSIS

Statistical intrusion analysis is the process of establishing a known footprint or baseline of a system's usage of such things as CPU (central processing unit) utilization, disk utilization, use of user rights, user logins, file and folder access over time, and analyzing the system for any deviation from the system's baseline or "normal" behavior. Most commercial IDSes reference an operating system's log files to establish a baseline. If you have used operating systems such as Windows NT or Windows 2000 before, the Windows Event Viewer should come to mind at this point. If you have never used an operating system before, you probably shouldn't attempt to take this exam.

NETWORK-BASED IDS (NIDS)

Network-based intrusion detection systems or *NIDS* are a combination of agent detection programs placed in certain bottled up areas of a network that gather network traffic that is from a NIC that is placed in promiscuous mode. The NIDS reads all incoming and outgoing packet header information and determines if the information or signatures in the header information match malicious code. If there is a positive match, or if misuse is detected, an alert is typically sent to a network specialist.

HOST-BASED IDS (HIDS)

Host-based intrusion detection systems, or *HIDS* for short, are IDS programs that are installed on single-server computers or client machines. HIDS only monitor intrusions on the individual machine they are installed on. There are two types of HIDS—those that monitor all traffic on the individual systems whether it is a dial-up or network connection, and those that use agents to search for threatening activity. Properly configured HIDS are designed to program activity, and detect buffer overflow and worm signature virus activity without wasting precious system resources such as CPU or disk processing power.

FALSE POSITIVES

A *false positive* is simply a report or an alert from an IDS that details something other than an attack. The IDS thinks the normal routine or program is an attack. Misuse or the overburdening of a system or network resource by an internal employee is an example of a false positive. It is important to plan what type of IDS will be used, and what an IDS will monitor, before implementing IDS solutions in order to reduce the amount of false positives.

FALSE NEGATIVES

A *false negative* error occurs when an IDS system completely misses legitimate misuse, or an attack on a system or resource, by mistaking it for normal activity. The IDS allows the program to carry on as if nothing has happened. This is very, very dangerous.

In conclusion, you need a well-organized plan to best protect your network, prepare for, identify, and respond to intrusions in general. You should consider the following items in your plan:

- Establish, document, and maintain management policies for intrusion detection and response.
- Secure network perimeters with firewalls, routers, switches, and strong antiviral programs and filters.
- Implement identity and authorization methods such as one-time passwords, Kerberos, RADIUS and TACACS, digital certificates, and smart cards.
- Implement security monitoring tools in order to identify that an intrusion has occurred.

HONEY POTS

Honey pots are used to attract hackers and crackers. A honey pot is basically an unprotected system with no applied patches, operating system updates, or firmware updates that is used to attract, trap, and identify possible attackers.

The main goal of a honey pot or mouse trap if you will, is to trap, track, and record the trails of a possible attacker. These logged or tracked movements can then later be used as evidence in possible legal proceedings if you press charges against someone who has damaged your data.

To set up a suitable honey pot or decoy, you should first build an FTP, DNS, or Web server outside of your normal DMZ safe zone (DMZ is described later). In other words, you want to protect your production servers, data, and IP addresses while you provide a suitable decoy or target. The key is to allow inbound traffic to a public IP address while restricting outbound traffic from the server. Setting up a firewall restricting outbound services from your honey pot server can accomplish this.

The following are important notes to remember when setting up a useful honey pot:

- Do not leave access to production data and server opens.
- Do not use strong passwords on the honey pot.
- Do not implement production user IDs and passwords on honey pots.

- Implement a real-time monitoring solution on your honey pot.
- Implement an alerting service on your honey pot so that you know when you are being attacked.

Setting up and maintaining a honey pot takes a lot of work. However, a honey pot is considered a very useful form of intrusion detection if properly instituted.

INCIDENT RESPONSE

Incident or *security incidence responses* are the measures taken in reaction to or as a result of a breach in network security. It is of utmost importance that any company or business providing a product or service has a structure in place in order to react to a security violation or threat to the well being of daily operations.

You should have a security response team and related documentation and procedures established before a critical situation arises. Unfortunately, the information world and its technology are becoming more and more complicated. Most companies seem totally interested in directing most resources toward direct profitability. They are not as interested in directing resources towards security, backups, and response. It takes money to have enough staff to implement, plan, document, and prepare for proper response and recovery. It seems like most companies direct their resources towards implementation and direct bottom line. Fair warning: do whatever it takes to implement a good, solid security response program that fits the needs of your particular network infrastructure.

A good, solid incident response procedure should be made up of the following items:

- Written procedures.
- Steps that will be implemented to correct, repair, or restore whatever has been damaged.
- Determination of who will be notified.
- Decision about how and when will they be notified.
- Sign off (in writing) that the plan was tested.

NETWORK COUNTERMEASURES

A *network countermeasure* is considered the implementation of an action, software, or the placement of a physical component in order to reduce the threat of intrusion or attack from outside influences. The best ways to protect your network and reduce the threat from outside your network perimeter is through the implementation of Virtual Private Networks (VPNs) and firewalls. Pay very close attention to the subjects discussed in this section. Chances are, you will be asked several questions on the real exam related to this subject matter.

VIRTUAL PRIVATE NETWORKS (VPNs)

As soon as the efficiency and security of LANs (local area networks) became evident to administrators, it became clear that these secure networks needed a farther reach. Until the advent of *VPN (virtual private networking)*, this involved the costly use of fiber-optic or ISDN leased lines. Unfortunately, the expense of maintaining a private WAN (wide area network) grows with the distance from point A to point B. So, once the Internet proved itself as a reliable way to exchange data, businesses naturally turned to it to extend their networks. The Internet, though, is as wide open as the big skies of Montana. A process was needed to harness the global reach of this massive TCP/IP network while ensuring secure, reliable, and quick communications. Enter VPN.

In a nutshell, a VPN is a private network routed through public lines. VPNs use virtual connections established over the Internet as opposed to a dedicated, leased-line approach. They are used to connect multiple LANs (site-to-site) as well as to facilitate communications with users in the field. For instance, once a mobile user connects to their ISP (Internet service provider) from any location, they can then be authenticated on the company network over a VPN. This type of remote access VPN is also called a *VPDN (virtual private dial-up network)*. It's fast, secure, scalable, and best of all: the intricacies are transparent to the user.

So, now that you have an idea of what VPN is and why it's used, let's talk about security. By nature, the study of security is a sensitive endeavor. If you have been hired to establish a secure network, you'd better be sure that's what your employer ends up with. So, how does this VPN achieve such security through the vast gulf that is the Internet? There are actually several ways VPN keeps your data secure but don't take my word for it. You need facts that are discussed in the following sections.

Firewalls

The first step in implementing a VPN is having a solid *firewall* installed. Firewalls are used to prevent access to private networks from the Internet or other interconnected networks. They can come in either hardware or software form, or a combination of the two. A firewall can be programmed to block traffic based on the type of packet trying to get through, which protocol or port is being used for the transmission, or a host of other user-defined rules. A firewall is an implementation of software, hardware, or a combination of both, specifically designed to keep unauthorized users, programs, and other threats from entering a computer system or network. A typical firewall analyses every packet of information that attempts to enter or exit a network or computer system. If the packet does meet the specifications implemented by the firewall, the packet or connection is denied access. Several implementations of firewall techniques are provided through the use of a packet filter, a proxy server, or an application or circuit gateway. (Firewalls will be discussed in more detail shortly.)

IPSec (Internet Protocol Security)

Using a group of protocols developed by the IETF (Internet Engineering Task Force), this method enables the secure transfer of packets at the IP layer. *IPSec* employs two encryption modes: *transport* and *tunnel*. Using the transport mode, only the data portion (or *payload*) of a packet is encrypted while the header remains unchanged. In tunnel mode, security is further enhanced because both the payload and header are encrypted. An IPSec-compliant device (router, firewall, and so on) is required on the receiving end to decrypt the packet. Sending and receiving devices must both obtain a common public key for this method to work, which is accomplished by using a protocol called Internet Security Association and Key Management Protocol/Oakley (ISAKMP/Oakley). ISAKMP/Oakley also enables the receiver to authenticate the sender using a digital certificate.

Encryption

Encryption is a way of sending data in a form that only the intended receiving computer can decrypt. Most encryption systems use one of the two following standard ways of accomplishing this: *symmetric-key* and *public-key*.

Symmetric-key Encryption

The method employs a single key to encode and decode messages. While simpler than the public-key approach, this method requires the sending and receiving computers to exchange the key securely first. Think of this approach as a secret code that both parties must have in order to decrypt messages. Public-key encryption gets around this problem by transmitting the public key to anyone who wants it while never transmitting the private key.

Public-key Encryption

Also called *asymmetric encryption*, this method employs both a public key and a private key. Only your computer knows the private key and the public key is handed to any computer that wishes to engage in secure communications. To decipher an encrypted message, the receiving computer must utilize both the public key provided by you and its own private key. For instance, if Computer A wishes to send a secure message to Computer B, it would use Computer B's public key to encode the message. Computer B then uses its own private key to decrypt it. Using this method, it's near impossible to discover the private key if all you have is the public key. This process was developed in 1976 by Whitfield Diffie and Martin Hellman and is sometimes referred to as Diffie-Hellman encryption. One public-key system, Pretty Good Privacy (PGP), is now widely used for secure transmissions over the Internet.

Tunneling

Another technology that enables VPNs to work is called *tunneling*. This process involves encapsulating a network protocol within a packet prior to transmission. To tunnel data, the three following types of protocols are used:

- **Carrier protocol:** The native protocol used by the network carrying the data.
- **Tunneling protocol:** The protocol (PPTP, IPSec, L2F, L2TP, or GRE) that the original data is encapsulated in.
- **Passenger protocol:** The protocol (IPX, NetBeui, or IP) native to the original data.

Tunneling enables the secure transmission of packets using protocols not supported by the Internet to be sent inside IP packets. For instance, a remote user sends a message from their laptop over the VPN. The original message was in the form of a NetBEUI packet, which is useless on the Internet. Once this packet is encapsulated in an IP packet, it can make its way through the Internet via PPP, for example, and arrive at its destination back in its original form.

In a VPDN, PPP is usually used to tunnel packets. A few other protocols developed in the image of PPP are L2TP (Layer 2 Tunneling Protocol), L2F (Layer 2 Forwarding), and PPTP (Point-to-Point Tunneling Protocol). In a site-to-site VPN, GRE (Generic Routing Encapsulation) or IPSec in tunnel mode are most often the tunneling methods of choice.

SECURE REMOTE PROCEDURE CALL (RPC)

Secure Remote Procedure Call (RPC) is essentially a protocol that is used to allow a client-side application program to execute or request a service from a server computer without being concerned with network intricacies or server procedures. The RPC protocol operates at the Transport and Application layers of the OSI reference model. RPC allows for multiple programs to be easily distributed and executed across a network environment.

FIREWALLS

As mentioned earlier, a firewall is typically a combination of hardware or software or both, placed between two networks in order to protect an internal network from outside influences. The three main implementations of firewalls are packet filter techniques, circuit gateways, and application proxies. Each of these are described next.

Packet Filtering

Packet sniffing programs and *network monitors* can capture and analyze network packets coming into or going out of a network. These tools can identify bad NICs as well as malicious activities and send alerts to security administrators. What most packet filters actually do is examine UDP and TCP ports and packet header information. Table 4.2 displays some of the best known TCP and UDP ports, and the upper-layer protocols that use them.

TABLE 4.2   Important TCP/UDP Ports

Type	TCP/UDP	Port Number
FTP	TCP	20, 21
Telnet	TCP	23
SMTP	TCP	25
SMTP\Trap	UDP	161\162
DNS	UDP	53
HTTP	TCP	80
NetBIOS	UDP	137-139

You should be very familiar with Table 4.2 for the exam. The exam will most likely expect you to know that there are 1024 well-known ports.

Packet filters identify good from bad packet information. However, the main disadvantage to most of the packet filtering programs available today is their inability to identify whether the packets were sent by a normal, innocent user or a threatening, vicious source. It is important to note that most network routers in use today implement some sort of packet filtering. This can effectively provide protection to a certain extent. Unfortunately, most routers do not have effective logging capability, which can allow certain attacks to go unnoticed. Remember, packet filters are considered to be a basic line or first generation of firewall defense. However, they can allow or disallow network access based on port- and or protocol-related information. The advantage of a packet filtering router is its ability to analyze and drop or pass packets quickly. The major disadvantage of this technology is that packets are not typically analyzed beyond source and destination address.

Packet filtering firewalls operate at the Network layer of the OSI reference model.

Circuit Gateways

The effectiveness and functionality of a *circuit gateway* lies between a packet filter and an application proxy. A circuit gateway operates at the Session layer of the OSI reference model. It is essentially a packet filter that relays packets for one host to another based on protocol and IP address. A circuit gateway forms a sort of tunnel through a firewall allowing two specified hosts to interact. The main advantage of a circuit gateway is its ability to log more information than a simple packet filter.

 An attacker who sends a bogus protocol package through an open port can easily fool circuit gateways and packet filters. Thus, another stronger level of firewall can be implemented that allows more protocol and port control to administrators. Enter the application proxy.

The major advantage of a circuit gateway firewall is that it supports NAT (Network Address Translation.) NAT will be discussed shortly.

Application Proxies

Applications proxies or *gateways* operate at the Application and Presentation layers of the OSI reference model. They are concerned with specific applications and actual data. The application proxy offers much more control than packet filters and circuit gateways by controlling or limiting user access from within the protocol itself. In other words, with an application proxy, administrators can actually control what information can be sent out or pulled into a network. Application proxies allow untrusted networks to be blocked and FTP traffic can be controlled. The major disadvantage of this type of firewall is the administrative overhead required to maintain this type of control. It is common practice today for most administrators to implement a combination of circuit and application technologies to manage and protect an internal network properly.

There are three important notes regarding application proxy firewalls:

- They must be configured for each separate service that is to be analyzed.
- They are intelligent enough to understand information contained within protocols.
- They monitor the state of every configured connection and have the ability to intercept or capture a properly configured channel.

Stateful Inspection Firewall

Stateful inspection firewalls have the ability to remember detailed information about packets that have previously passed through them. Then, they are able to compare and analyze this information and decide whether to let certain packets through the firewall. In other words, a stateful firewall can compare incoming requests to outbound messages and see if there is a relationship between the two. If not, the firewall can block the incoming request.

Stateful firewalls provide better overall analysis than most other firewall types such as packet filters. They have the ability to compare packet information over time and provide for a much more effective firewall solution.

 There is a possibility that the exam might display a routing configuration and ask you what the administrator is trying to protect the network from. In other words, what type of an attack is the administrator attempting to circumvent?

Network Address Translation (NAT)

Network Address Translation (NAT) is an Internet standard. It is most often used with routers to provide firewall security by hiding an internal private network's range of IP addresses from outside networks. What NAT actually does is translate a given set of internal IP addresses to a separate set of IP addresses that can be seen and accessed outside of a local area network. The translation of internal to external and external to internal IPs provides security and a sort of authentication service to requests.

Another benefit of using NAT is that many internal IP addresses can be translated to use one external address. This allows a company to conserve global IPs and use one address to connect outside of the network. For example, a company could combine or map the IP addresses of several ISDN connections into one connection to the Internet. It is possible for administrators to use NAT to map in the following fashions:

- Map a local network IP address statistically to a single global IP address.
- Map a local Network IP address to a pool of global IP addresses owned by a company.
- Map a single global IP address to a pool of local network IP addresses.
- Map a local network IP address and TCP port address to a global IP address or global pool if IP addresses.

DEMILITARIZED ZONE (DMZ)

A *Demilitarized Zone (DMZ)* is a neutral area between an internal network and the Internet that typically contains one host system or a small network of systems. The DMZ sits between a private and a public network and can be made up of one or several systems that house Web pages and non-critical company data that can be accessed from outside an intranet or LAN. In other words, a DMZ is located behind a firewall and can be seen from the Internet. If the Web pages or non-critical data are destroyed or lost, it is usually not a major catastrophe in a DMZ based on the fact that the applications that support the Web pages and critical data are stored within the local network.

A DMZ host can accept Internet requests from SMTP, FTP, HTTP, or name resolution services such as DNS. The real benefit from using a DMZ is simply the ability to provide certain information to outside sources without allowing unauthorized access to critical internal services, information, and programs. DMZ capabilities and products are available from such manufactures as Cisco Systems.

Important servers that provide services such as FTP, SMTP, HTTP, SNMP, Telnet, and DNS should all reside in a DMZ.

DEVICES

Network devices such as routers, switches, bridges, and hubs connect computing systems and networks. They are responsible for productive network functionality, backbone support, and the proper forwarding of information to other networks. In short, without them, there would be little or no network connectivity and minimal network security at best. In order to grasp fully the network infrastructure security concepts, it is essential that you have a general understanding of how these devices work.

ROUTERS

A *router* is a network device that is used to connect networks. They are most often used to connect LANs. A router uses packet header information in combination with a sort of stored database known as a *routing table* to determine the best route to use with the forwarding of packets to other networks or subnets (subnetworks.)

Routers use specialized protocols such as ICMP (Internet Control Message Protocol), OSPF (Open Shortest Path First), and RIP (Routing Information Protocol) to communicate with each other and carry out their advanced functions. ICMP is the most common protocol used by modern day routers. It allows for actual packet information to be read and provides support for packet error correction.

Routers connect different network segments. However, unlike bridges, routers do not use a computer's MAC address to forward information. Instead, a router operates at the network layer of the OSI reference model and has the ability to forward information based on a network or individual computer's TCP/IP address. This allows a router to connect entirely separate networks and filter information to the proper network or network segment. In other words, a router has the ability to send a request to a specific location without broadcasting to all of the other computer nodes on a network or network segment.

Routers are very intelligent. As mentioned earlier, they maintain sophisticated routing tables and have the ability to remember previous connections that where used as pathways from one computer node to another.

Here are some important points to remember concerning routers:

- They provide filtering of packets and reduce broadcast storms.
- They can segment networks into smaller and more manageable pieces.

- They provide a network security layer between separate networks, functioning as firewalls.
- They connect LAN segments that use the same or different protocols.

SWITCHES

A *switch* is a network device similar to a router that chooses certain paths or routes in a network on which to send data. A switch is not a router although a switch can contain router functionality. Most modern day switches can operate at both the Data Link and Network layers of the OSI reference model. A switch that has the ability to operate at the Network layer is known as a layer 3 or IP switch. Switches can connect networks and subnetworks comprised of the same or different cable types. They can send units of data (packets) faster than most routers based on digital packet-switching technology. Switches connect LAN segments that typically use the same protocol.

Asynchronous Transfer Mode (ATM) Switches

ATM switches use cell relay switching technology that combines both conventional circuit and packet switching technologies. This results in a high-speed switching process that is well suited to support today's video and audio streaming technologies.

BRIDGES

Bridges are hardware devices that operate at the MAC sublayer of the OSI reference model's Data Link layer. Bridges are used to segment or separate LANs. Separating a larger network into smaller manageable segments can improve network performance and provide a way to isolate network bottlenecks.

A bridge reads the MAC hardware address that is stored in the NIC of every computer or node installed on either side of the bridge. The bridge knows where all of the computers are on the network and can forward information to a particular computer by the use of its NIC MAC address. Let's say, you are sitting at your computer that resides on network segment number 1. You want to send Brian a Word document and his computer is located on network segment number 2. There is a bridge that separates you on network segment number 1 from Brian on network segment number 2. The bridge can identify both of your computers by their respective network interface card's MAC address. So, when you send a Word document to Brian, it is forwarded to his network segment by the use of the bridge.

Bridges can provide the following services:

- Reduce network traffic as a result of too many computers being attached to a network.
- Connect different types of media connections such as coaxial cable and twisted pair cable.
- Expand the length of a network segment.
- Connect different network typologies such as Token Ring and Ethernet.

Although bridges serve their primary purpose, they are limited in their capabilities. If a destination's MAC address is not found in a bridge's internal table, the bridge will proliferate or broadcast (pass traffic) to all network segments. This can result in a broadcast storm that can slow or take down a network.

As networks grew larger, the demand for a more intelligent device that could handle more attached computer nodes and direct network traffic in more efficient manner increased. The router was technology's answer to this demand.

HUBS

As mentioned earlier in this chapter, a *hub* is a network device that acts as a central point used to connect computers. In network terms, a hub is a simple connection device that sends all data packets to all connected systems. A basic hub operates at the Physical layer of the OSI reference model.

TELECOM/PBX (TELECOMMUNICATIONS/PRIVATE BRANCH EXCHANGE)

A *telecom/PBX* or just plain *PBX* is a telephone-switching system inside an enterprise that allows calls to be shared or switched to various numbers assigned within the enterprise. Callers inside of the enterprise also share external PBX lines for outbound or outside calling. A PBX offers the flexibility of inside calling numbers that are typically only three to four digits in length. The newest PBX technology is called Centrex. With Centrex, all telephone switching takes place at the local phone company, as opposed to switching at the customer's premises.

MODEMS

There are many types of modems available on the market today. Popular modems types include cable, wireless, Digital Subscriber Line (DSL), and analog dial-up modems. *Modem* is an abbreviation for *Modulate Demodulate*. A computer sends data from the CPU to a modem in digital format. For a standard analog modem, the modem (modulator) coverts digital data to analog format that can be sent over a POTS (Plain Old Telephone Service) line. When the analog signal reaches the receiving modem, it is converted back to digital format that can be understood by the receiving computer.

There are many ways your system can be attacked. Today, most systems are attacked by computer viruses through operating system and application flaws. The simplest way to safeguard your system if you use a modem is to remove access to your files and folders. Assuming that you are using Windows, ensure that you do not have file and print sharing enabled. This can accomplished using the Network Neighborhood Icon located in the Control Panel applet.

You should also have a good antivirus program and make use of a personal firewall system. Other than that, once again, **back up your important data!**

SECURITY BASELINES

Of all the topics you need that to study regarding the Security+ exam as well as other security certification available today, you will find security baselines to be the most general and undefined subject of them all. This is simply because there is no particular set of standards or defined baseline rules that can be applied to every possible network infrastructure in use. Simply put, if you are responsible for administrating security in a networked environment, it is important for you to set your own security baselines based on your network environment.

Here are some good general guidelines to follow when creating a security baseline for your network:

- Evaluate your company's current processes, business plans, technical environment, and current security structure.
- Identify your company's security risk. This can be accomplished with a fundamental risk analysis and a good network security analysis scanner.
- Plan and set up a strong authentication method for your network. Implement encryption methods such as use of public/private keys and utilize built-in operating system security practices such as file-level security, passwords, and policies

- Plan and provide protection for remote users through the use of VPNs, firewalls, and extranets
- Create a *choke point* (a single entry and exit point through which data passes into and out of your network). This can be done with a firewall. It will provide one area to protect, monitor, and log.
- Define a security policy.
- Secure all resources and services.
- Perform testing, evaluation, and logging.
- Create multiple network segments.
- Segment operating systems from data. Partition hard drives so that the operating system is separate from data.

 Do not take this section lightly. There is a good possibility that the general guidelines just mentioned contain several answers to questions you will face on the exam. For example, what should you do first in preparation for avoiding potential risk? The exam is most likely going to drill you with such questions. Use your technical knowledge and common sense to answer the easier exam questions.

In order to create a structured network security baseline, you have to know the location of your vulnerabilities. In other words, you need a tool that identifies your network's weaknesses. There are some great tools available that will help network security personnel assess weaknesses and create security baselines. One of the best on the market is Enterprise Security Manager by Symantec. For more information on these as well as other excellent security software packages, visit the Symantec Web site at *http://www.symantec.com/product/*.

TEST TIPS

It has been stated already that security is a very broad topic. In order to prepare yourself for the Security+ exam and the plethora of possible questions you might encounter, you should focus your test preparation study on the following network security-related Test Tips as well as the review questions at the end of the chapter.

These Test Tips serve primarily as a review of the chapter. However, you might notice that some of the tips have not been discussed. Be prepared and learn to be surprised by the unexpected. The real exam will show you no mercy. Know these tips inside and out!

√ *Multiplexing* is the combining of data channels over a single transmission line.

√ *DNS (Domain Name Servers)* resolve fully qualified domain names (or host names) to IP addresses. For example, *Microsoft.com* is a domain name. A properly configured DNS could resolve *Microsoft.com* to the IP address 207.46.129.180.

√ A *bastion host* is gateway or firewall that protects an internal network from external networks. Simply put, a bastion host is a system setup on an internal network that screens for possible attacks aimed at a particular internal network.

√ *ATM (Asynchronous Transfer Mode)* is a dedicated switching technology that transmits data in fixed-length, 53-byte units called *cells*. ATM is well suited for the transmission of audio and video and is said to be the answer to the low bandwidth problems that face Internet users.

√ A *Demilitarized Zone (DMZ)* is a neutral area between an internal network and the Internet that typically contains one host system or a small network of systems.

√ *Network Address Translation (NAT)* is an Internet standard most often used with routers to provide firewall security by hiding an internal private network's range of IP addresses from outside networks.

√ An *FDDI ring* is typically composed of two fiber-optic Token Rings. An outside ring is used for the primary transport of data and an inside ring acts as a backup if the primary ring fails. An FDDI ring is redundant and somewhat more secure than star, bus, or traditional ring topologies.

√ The seven layers of the OSI reference model from layer 1 to layer 7 are Physical, Data Link, Network, Transport, Session, Presentation, and Application. Just remember "Programmers Do Not Throw Sausage Pizza Away."

√ *Secure Remote Procedure Call (RPC)* is a protocol that is used to allow a client-side application program to execute or request a service from a server computer without being concerned with network intricacies or server procedures.

√ *IPSec* employs two encryption modes: *transport* and *tunnel*. Using the transport mode, only the data portion (or *payload*) of a packet is encrypted while the header remains unchanged. In tunnel mode, security is further enhanced because both the payload and header are encrypted. IPSec offers security services such as connectionless integrity, data origin authentication, and confidentiality.

√ *Frame relay* is also much faster than X.25 and can take advantage of T1 (1.544 Mbps) and T3 (Mbps) speeds. Frame relay uses public switched WANs that can redirect packets if a segment goes bad.

√ A *circuit gateway* is a packet filter that relays packets from one host to another based on the protocol and IP address. A circuit gateway forms a sort of tunnel through a firewall allowing two specified hosts to interact.

√ The term, *transparency*, is used in network security lingo to describe how intrusive a network countermeasure, such as a firewall, is to a user. For example, a packet filter is more transparent to a user than an application proxy. In other words, users will typically not be aware that a router is filtering their data packets. However, if an application gateway is used, users will have to authenticate with the firewall or configure their applications to authenticate through the firewall.

√ *Point-to-Point Tunneling protocol (PPTP)* allows a virtual private network (VPN) to be created using the Internet. PPTP is essentially a set of communication rules that allow the boundaries of private networks to be extended. PPTP has in many cases eliminated the need for companies to use expensive, dedicated leased lines to expand the privacy of their networks.

√ Leased-line speeds are as follows:

- **DS-0 (Digital Signal Level 0):** One channel transmits 64KBps on T1 line.
- **DS-1 (Digital Signal Level 1):** Transmits 1.544MBps on T1 line.
- **DS-3 (Digital Signal Level 3):** Transmits 44.736 MBps on a T3 line.

√ *CAT5 UTP* is also referred to *100BaseT* or *100BaseTX*. It carries a data signal 100 meters or approximately 328 feet. It is the most popular UTP cable in use today.

√ *Application proxies* or *gateways* are concerned more with specific applications and actual data. The application proxy offers much more control than packet filters and circuit gateways by controlling or limiting user access from within the protocol itself. In other words, with an application proxy, administrators can actually control what information can be sent out of or pulled into a network.

√ *CHAP* uses a secret one-way hash value that is generated by the requester and sent to the server.

√ *SMTP (Simple Mail Transfer Protocol)* is an unsafe protocol used to send e-mail messages between mail servers. SMTP was not originally developed to protect against e-mail and e-mail server attacks. The best way to protect your e-mail server and e-mail in general is to scan and filter all messages and secure each e-mail message with encryption.

√ *SNMP (Simple Network Management Protocol)* is an unsafe network management protocol that allows the use of clear text passwords. SNMP traffic should be filtered at the firewall.

√ A *multihomed server* (a system with two NICs) can be configured as a firewall by enabling IP forwarding and building a *Routing Information Table (RIT)*.

√ Devices such as routers, hubs, and switches are a single point of failure. Each of these devices should be protected with a *UPS (Uninterruptible Power Supply)* in the event of power surges, spikes, and brownouts.

√ The four primary types of firewall architectures are as follows:

- **Packet filter:** A *packet filter router* uses an ACL (Access Control List). It is the oldest of the mentioned architectures. It separates a private network from a public network.
- **Screened host:** This firewall architecture combines a bastion host and a packet filter firewall, which requires the intruder to get by two separate systems in order to reach an internal network. This is more secure than a traditional packet filtering firewall.
- **Dual-homed host:** Straight to the point, this is a system with two NICs. One NIC supports access to a private network and the other is for a public network. This acts as a filter and is also known as a *multihomed bastion host*.
- **Screened subnet:** This firewall architecture combines the security of two packet filters and a bastion host. This is the most secure of the firewall architectures and requires high overhead. This overhead is realized in high-maintenance requirements.

√ *VoIP (Voice over IP)* technology is essentially the delivery of voice in digital packets over IP networks. This technology is generally less expensive than traditional circuit switching of voice using PSTN (Public Switched Telephone Network). VoIP is a rapidly growing technology that offers security and quality of service.

√ *Security services* are a combination of security techniques, files, policies, and procedures. The following six security services are defined by OSI communication standards in order to provide secure communications:

- Authentication
- Access control
- Data confidentiality
- Data integrity
- Non-repudiation
- Monitoring and logging

√ In order for packet-switching networks to work properly, packets must contain the network address of the sending system as well as that of the destination system.

√ When the functionality of devices such as a network bridge and a network router are combined, the result is a device known as a *brouter*.

√ Most communication takes place at the Data Link layer of the OSI reference model.

√ An *extranet* is part of a private network (intranet) that is extended to customers, vendors, suppliers, and possibly other remote users. Most extranets use tunneling to connect multiple intranetwork connections to an extranet.

CHAPTER SUMMARY

If you found the information in this chapter a bit overwhelming, you are not alone. This is considered the second most difficult chapter in this book. If you found the information in the chapter fundamental, you should consider broadening your knowledge of the topics discussed with such tools as the Internet and reputable security books that specialize on the individual items discussed in this chapter.

At this point, you should know the following at the very least:

- Understand the seven layers of the OSI reference model and what actually takes place at each level.
- Be able to identify and describe various network topologies and the cabling used for each.

- Be able to define security-related protocols and have a basic understanding of the required and unnecessary services.
- Understand the most popular network countermeasures that can be implemented to protect a network from possible intrusion and attacks.
- Have a basic understanding of network devices such as repeaters, hubs, routers, and bridges.
- Understand media security procedures and security baselines.

If you want to pass this newly developed exam and acquire this important certification that is considered in high demand, the time to focus is now! Please use the knowledge you have gained from this chapter to best answer the review questions that follow.

REVIEW QUESTIONS

1. **Name the minimum category cable type that can be used to support 10BaseT.**
   - A. Category3.
   - B. RJ-11.
   - C. Catagory4.
   - D. Category5.

   Correct answer = A

   The minimum cable type needed to support 10BaseT is Category3. An RJ-11 phone connector is used for earlier categories of UTP to connect a modem to a typical phone jack or your phone to a phone jack. Catagory4 and Category5 cable types do support 10Baset Ethernet specifications. However, they are not the minimum category type needed to support 10BaseT.

2. **Which IEEE specification is concerned primarily with security?**
   - A. 802.3.
   - B. 802.5.
   - C. 802.10.
   - D. 802.11.

   Correct answer = C

802.3 is concerned with Carrier-Sense Multiple Access with Collision detection in local area Ethernet networks. The 802.5 specification is for Token Ring LANs. 802.11 is an IEEE specification for wireless communications.

3. **Which type of network cabling is the most difficult to tap into and considered the most secure?**

   ○ A. Shielded twisted pair.
   ○ B. Twisted pair.
   ○ C. Coax.
   ○ D. Fiber.

   Correct answer = D

   Fiber-optic cable is much more difficult to tap into than other types of network cable. Special equipment and skilled hands are required to carry out such a task. This is not the case with twisted pair, shielded twist pair, and coaxial types of cable.

4. **Which statement is true regarding firewalls?**

   ○ A. They will protect your internal network from a virus that resides on a workstation on your internal network.
   ○ B. If properly configured, a firewall will protect an internal network from an external network.
   ○ C. Firewalls are used to protect your internal network from unauthorized external access through dial-up modem connections.
   ○ D. Firewalls are used to protect internal server-room computers from external natural disasters.

   Correct answer = B

   Properly configured firewalls will protect an internal network from an external network. An antivirus program and updated operating system service packs would be used to protect your internal network from a virus that resides on workstations on your internal network. Firewalls do not provide protection through dial-up modem connections nor do they protect against natural disasters.

5. **Which of the following describe an FDDI ring? (Choose three)**
   □ A. Offer up to 100Mbps data transmission speeds.
   □ B. Are composed of dual rings with data traveling in opposite directions.
   □ C. Can serve as a network backbone.
   □ D. FDDI is the easiest topology to tap into.
   □ E. FDDI uses CSMA/CD technology.

   Correct answers = A, B, and C

   Answers A, B, and C are all true statements regarding FDDI rings. Option D is a sort of a trick. FDDI uses fiber cable, which as you are already aware is more difficult to tap into than most other cable types. FDDI uses token passing technology not CSMA/CD (Carrier-Sense Multiple Access with Collision Detection.) CSMA/CD is used in Ethernet networks.

6. **Which of the following choices forms a sort of tunnel for two hosts to communicate?**
   ○ A. Packet filter.
   ○ B. Circuit gateway.
   ○ C. Application proxy.
   ○ D. FDDI.

   Correct answer = B

   A circuit gateway forms a sort of tunnel through a firewall allowing two specified hosts to interact. Packet filters examine UDP, TCP ports, and packet header information. They can identify good from bad packet information. Application proxies (or gateways) are concerned more with specific applications and actual data. FDDI is a network topology standard that utilizes dual fiber-optic rings.

7. **Of the following protocols, which protocol uses a one-way hash function to assist with the authentication process?**
   ○ A. Password Authentication Protocol (PAP).
   ○ B. Challenge-Handshake Authentication Protocol (CHAP).
   ○ C. Point-To-Point Protocol (PPP).
   ○ D. Serial Line Internet Protocol (SLIP).

   Correct answer = B

CHAP uses a secret one-way hash value that is generated by the requester and sent to the server. PAP is a basic type of authentication where a username and password are transmitted unencrypted across a network to an authenticating host. PPP is a full-duplex serial communication protocol that operates at the Data Link layer of the OSI reference model. Serial Line Internet Protocol (SLIP) is an older TCP/IP communications protocol used to connect two computer systems.

8. **Which of the following practices should be followed in order to harden an internal network from outside influences? (Choose four)**

   ☐ A. Apply the most recent manufacturer's hot-fixes and service packs.

   ☐ B. Block TCP port 139 and UDP ports 137 and 138.

   ☐ C. Disable (stop) all services and unbind all protocols on all internal network workstations.

   ☐ D. Enable a strong password structure.

   ☐ E. Enable auditing.

   Correct answers = A, B, D, and E

   The only incorrect answer to this question is C. If you chose answer C, please read the entire chapter again. If you disable (stop) all services and unbind all protocols on all internal network workstations, your workstations will not connect to a network and your operating system will not function.

9. **Which of the following protocols is used to map or "resolve" an IP address to a system's physical hardware address?**

   ○ A. HTTPS.

   ○ B. SDLC.

   ○ C. HDLC.

   ○ D. ARP.

   Correct answer = D

   Address Resolution Protocol (ARP) is a protocol used to map an IP (Internet Protocol) address at the network layer of the OSI model to a physical hardware address at the MAC (Media access Control) sublayer. HyperText Transport Protocol Secure (HTTPS) is a secure protocol used to transmit messages over the Internet. SDLC is based on a primary/secondary communications model where a secure connection is established between a mainframe (host) and a client. High Level Data Link Control (HDLC) is a transmission protocol that operates at the Data Link layer (layer 2) of the OSI model.

10. **Which of the following provides a type of firewall by hiding internal IP addresses from outside networks?**

    ○ A. ARP.
    ○ B. NAT.
    ○ C. PAP.
    ○ D. RPC.

    Correct answer = B

    Network Address Translation (NAT) is an Internet standard most often used with routers to provide firewall security by hiding an internal private networks range of IP addresses from outside networks. Address Resolution Protocol (ARP) is a protocol used to map an IP (Internet Protocol) address at the Network layer of the OSI mode, to a physical hardware address at the MAC (Media access Control) sublayer. Password Authentication Protocol (PAP) is a basic type of authentication where a username and password are transmitted unencrypted across a network to an authenticating host. Secure Remote Procedure Call (RPC) is essentially a protocol that is used to allow a client-side application program to execute or request a service from a server computer without being concerned with network intricacies or server procedures.

11. **Which device does not forward all broadcast traffic and has the ability to forward data packets to other networks based on IP address information?**

    ○ A. Bridge.
    ○ B. Router.
    ○ C. Hub.
    ○ D. Repeater.

    Correct answer = B

    A router operates at the Network layer of the OSI reference model and has the ability to forward information based on a network or individual computer's TCP/IP address. A router has the ability to filter out broadcast traffic. Bridges are limited in their capabilities. They forward packet information based on MAC addresses. A bridge proliferates or broadcasts (passes traffic) to all network segments. This can often result in a broadcast storm that can slow or take down a network. A hub is a simple connection device that sends all data packets to all connected systems. A repeater is used to boost or regenerate the signals placed on a 10base2 or bus network. Adding a repeater to your bus can extend the length of your entire bus network.

12. **Which TCP port is HTTP associated with?**

    ○ A. 21.
    ○ B. 25.
    ○ C. 53.
    ○ D. 80.

    Correct answer = D

    HTTP traffic uses TCP port 80. FTP uses TCP port 21. SMTP uses TCP port 25. DNS uses UDP port 53. Please refer to Table 4.2.

13. **Which intrusion detection device is used to lure and trap possible network attackers?**

    ○ A. False positive.
    ○ B. Honey pot.
    ○ C. False negative.
    ○ D. DMZ.

    Correct answer = B

    The main goal of a honey pot or "mouse trap" if you will, is to trap, track, and record the trails of a possible attacker. A false positive is simply a report or an alert from an IDS that details something other than an attack. A false negative error occurs when an IDS completely misses legitimate misuse of or an attack on a system or resource mistaking it for normal activity. A Demilitarized Zone (DMZ) is a neutral area between an internal network and the Internet that typically contains one host system or a small network of systems.

14. **What is the name used when access to part of an internal network is granted to outside venders and customers?**

    ○ A. Ethernet.
    ○ B. Intranet.

○ C. Ignorantnet .

○ D. Extranet.

Correct answer = D

When part of an internal network or intranet has been accessible to outside sources, that part of the internal network is referred to as an extranet. Ethernet is a LAN architecture technology developed by Xerox that supports CSMA/CD. You should have learned this as a prerequisite to studying for Security+. An intranet is a network that is considered private and separate from the outside world. It exists to connect the workings of an internal network. Ignorantnet is an invalid selection.

15. **Which of the following is a small plastic device that can be used for storage, authentication, or memory purposes?**

○ A. CD-R.

○ B. Smart card.

○ C. Intelligent token.

○ D. Tape cartridge.

Correct answer = B

A smart card is a small plastic card that contains a microchip. It can be used for data storage and memory purposes as well as a security authentication device. A CD-R is an optical storage disk capable of storing large amounts of data. Intelligent token is an invalid selection. Although a tape cartridge is often used for data storage, it is not an authentication device nor is it used for memory purposes.

16. **How long are IPv6 addresses?**

○ A. 16 bits.

○ B. 32 bits.

○ C. 64 bits.

○ D. 128 bits.

Correct answer = D

IPv6 allows for IP addresses to be lengthened from the IPv4 limitation of 32 bits to 128 bits. IPv6 also allows for better authentication, privacy, and improved data delivery assurance. All other choices are invalid.

17. **Which statement best describes a DMZ?**
    ○ A. Used to connect two pieces of cable.
    ○ B. A DMZ should never be exposed to the Internet.
    ○ C. Located behind a firewall and can be seen from the Internet.
    ○ D. Located in front of firewall and can be accessed from the Internet.

    Correct answer = C

    The DMZ sits between a private and a public network and can be made up of one or several systems that house Web pages and non-critical company data that can be accessed from outside an intranet or LAN. BNC and BNC barrel connectors are typically used to attach or connect a bus cable to a device or connect one piece of the bus cable to another. Answers B and D are invalid.

18. **How long are IPv6 addresses?**
    ○ A. 16 bits.
    ○ B. 32 bits.
    ○ C. 64 bits.
    ○ D. 128 bits.

    Correct answer = D

    IPv6 allows for IP addresses to be lengthened from the IPv4 limitation of 32 bits to 128 bits. IPv6 also allows for better authentication, privacy, and improved data delivery assurance. All other choices are invalid.

19. **FTP uses which TCP ports?**
    ○ A. 25.
    ○ B. 20, 21.
    ○ C. 23.
    ○ D. 80.

    Correct answer = B

    FTP uses TCP ports 20 and 21. SMTP uses TCP port 25. Telnet uses TCP port 23. HTTP uses TCP port 80. You'd better know this for any security exam!

REFERENCES

http://standards.ieee.org/ is the Institute of Electrical and Electronic Engineers (IEEE) Web site where the IEEE Standards can be explored in detail.

http://www.symantec.com/product/ is the Symantec Web site that offers many of the world's leading antivirus and Internet security products and services.

BASICS OF CRYPTOGRAPHY

*C*ryptography entails the use of a wide array of hardware and software mechanisms to mask (or encrypt) information into an illegible format. This is done in order to transport sensitive information over insecure networks such as the Internet or simply to store data in a safe manner. Most cryptographic systems also provide a method of decrypting the information upon retrieval. As we will discuss, these systems generally use one of two methods to accomplish this transformation: *private-key* systems (symmetric), which employ a single secret key, and *public-key* systems (asymmetric) that use a combination of a public-key **and** a private key. The processes used to disguise information are generally characterized by their varying degrees of complexity.

CONCEPTS OF USING CRYPTOGRAPHY

Some cryptographic methods are actually designed to be irreversible, such as *hash* encryption. Hash encryption techniques provide a method of storing an encrypted representation (or fingerprint) of a data block. This is useful when you don't actually want to store the data in question, as with passwords. For example, on a UNIX system, a hash of your password is the only stored version that is available to the system. The password cannot be recovered from its hash. Rather, you enter your password and the system applies the hash algorithm to your entry. The result is compared to the stored hash and access is granted only if there is a match. This makes it theoretically impossible to break into the system and discover passwords.

As far as reversible encryption processes are concerned, few developers will claim that their product is hack-proof. Effectively, one of the goals of cryptography is to make it so hard to get to the encrypted data, the amount of time spent cracking the code is not worth the prize. This is the theory, at least. However, because the persistence of hackers continues to amaze us security professionals of goodwill, a plethora of techniques is now available that aid in the fight to protect sensitive data. Victory at the front lines of the Security+ examination will require your knowledge of these techniques and what makes some stronger than others. You should also have a full understanding of the key theoretical concepts of using cryptography, which include those discussed in the following sections.

CONFIDENTIALITY

This term couldn't be more straightforward. *Confidentiality* is simply the process of keeping data hidden from unauthorized persons. It is the main reason that we encrypt data in the first place. Obviously, there are endless reasons to keep data confidential, ranging from combat soldiers exchanging information on the battlefield without giving away their positions, to testing centers receiving the contents of your exam over the Internet without you getting a copy. In the latter example, confidentiality is really protecting you, the test taker. If the confidentiality of exam contents were repeatedly compromised, the value of certification would be lessened, the job market would be flooded with "security professionals," employers would not recognize certification as an asset, and you would be hard-pressed to find a job.

INTEGRITY

Guarding the *integrity* of data involves keeping it from being altered while in transit or while being stored. Following our last example, imagine an individual who wished failure upon you regarding your examination. If this malicious person could infiltrate the test data on its way to the testing center, they could mix up the answers and then send it back on its way. After reading this book, you would naturally select all the correct answers, still fail the exam, have no idea why, and again, you would be hard-pressed to find a job. This is a most undesirable scenario and it's important to understand that encryption alone will not protect data integrity. For this reason, many encryption programs have built-in measures to check data integrity. Although closely related, confidentiality and integrity have different objectives. Be aware of the differences.

One tool used in the verification of data integrity is the *message digest*. A message digest is essentially a fingerprint of a chunk of data. For example, the MD5 hashing algorithm can create a unique fingerprint for a file based on its contents. The process is irreversible and the contents of the original file can't be determined from its fingerprint. Upon receipt of the file, the algorithm is reapplied and the resultant fingerprint is compared to a known good one, which might be stored on a server at the original location of the file. If it's a match, it's safe to say that the file's contents were not altered in transit and its integrity is intact. *Digital Signature* (DS) methods, which we'll discuss shortly, also include integrity verification features.

 Message digests provide a method of verifying data integrity.

NON-REPUDIATION

Repudiation is the denial of involvement in an agreement—more specifically, an agreement that involves the transfer of money. When you go out to dinner and charge the bill on your credit card, the restaurant requires you to sign their copy of your receipt. This is done to protect them if you don't remember eating out that night and you call to dispute the charge. With your signature on file, it's very simple for the restaurant owner to prove to the credit card company that you did indeed eat there. This is a *non-repudiation* security measure built into the credit card system. This procedure also protects you if your card is stolen, giving you the ability to prove that your signature was forged. It's simple and sweet. In extreme cases, either party involved in a dispute might call upon an expert to prove or disprove the *authenticity* of a signature. Digital Signatures (DS) and time stamps are among the methods used to provide non-repudiation in the realm of electronic documents.

 Digital Signatures provide non-repudiation.

AUTHENTICATION

As we learned in Chapter 2, *authentication* is the process of proving that you are who you say you are. Whether you are logging onto a network, signing an encrypted e-mail, or withdrawing cash at the ATM, there must always be an authentication scheme in place to provide this facet of security. Sometimes this process takes place without any user input, as with Secure Socket Layer (SSL) transmissions. The server authenticates **itself** during SSL-driven communications. In a similar fashion, your browser, on its own, generates a temporary key for each session.

The methods and organizational systems directly related to cryptography that provide authentication include passwords, *Key Distribution Centers (KDC)*, *Digital Signatures (DS)*, and *Certification Authorities* (CA). Remember, you need to have a handle on the differences between *authentication* and *authorization*. They look and sound alike but they have distinctly different purposes.

DIGITAL SIGNATURE (DS)

The elegance and effectiveness of the handwritten signature has kept it a worldwide standard for ages. Despite all of our technology, we still rely on it today. Signing an electronic message was not always so simple. It became apparent that a digital equivalent of the signature was needed and the *Digital Signature* (DS) was born.

It is within the public-key encryption scenario that a DS is generally applied to an electronic message. The purpose of the DS is to verify that a message is truly from the sender noted in the message. A DS can also provide measures to ensure the integrity of a message. Although the DS process is usually handled behind the scenes by encryption software, it has a few steps that should be understood in order to sign and verify electronic documents properly. Now, let's examine a brief overview of signing a document digitally.

First, the sender's software uses a hashing algorithm to create a finger-print (or message digest) of the message contents.

 If the data being transmitted is changed in any way, its fingerprint will be changed as well.

The message digest, personal information about the sender, and possibly a time stamp, are then encrypted using the sender's private key, creating the DS. The resultant ciphertext string (the DS) is appended to the message and everything is encrypted with the recipient's public key. The message is now signed, encrypted, and sent on its way.

Upon retrieval, the recipient's encryption software will decrypt the message using their respective private key, exposing the DS and the original plain text message. The extracted DS is decrypted with the sender's public key, revealing the fingerprint of the original message. Using the same algorithm as the sender, the recipient's software will fingerprint (or hash) the original, plain text message. This fingerprint is compared to the one extracted from the message, and if it matches, you can be sure who it's from and that its contents have not been altered. At this point, you have verified the *signer* of the document and the *integrity* of its contents. This process also provides a non-repudiation security measure that discourages the denial of involvement in a digital transaction.

There are a few standards in place that govern the creation of digital signatures. The DSS, or *Digital Signature Standard*, uses the *Digital Signature Algorithm* (DSA) developed by the National Security Agency (NSA) and is used for the creation of digital signatures. Specified in Federal Information Processing Standard (FIPS) 186, DSS is the U.S. government standard algorithm used for signing electronic data. The National Institute of Standards and Technology (NIST) brought DSS to the public in 1994.

That's all fine and good, right? The DS is very effective but how do you really know that when your software extracts a digital signature, the signature in question is truly *authentic*? The answer lies within the *Certification Authority* (CA), which acts as the digital counterpart to the handwriting analyst. CAs use several techniques to guarantee the authenticity of a DS. Stay tuned; we'll discuss CAs in greater detail later in the chapter.

ACCESS CONTROL

Access control is mentioned again here to emphasize its role as one of the key concepts of using cryptography and its importance in the study of security as a whole. (Review Chapter 2 to obtain more information about access control.) As you are learning, this study demands skills that cover a vast ocean of topics and disciplines, extending far beyond the reach of this book and the test it's preparing you for. By now, you should be starting to get a clearer picture of the more pertinent subject matter that you're likely to see in the exam.

You'll recall that access control is a term for systems in place that manage access to resources. Remember Windows 2000/NT Discretionary Access Control (DAC) and the Access Control List (ACL) that it uses? Windows 2000 access control works in conjunction with the Kerberos protocol. In turn, this secure protocol takes advantage of different algorithms to deliver secure logon capabilities. The algorithms used in Kerberos include DES and RC4. RC4 is the default choice while DES support is available to enable interoperability with other applications.

What exactly is an algorithm? Good question—we're about to learn about algorithms.

ALGORITHMS

The *algorithm* is essentially the mathematical building block with which encryption techniques are constructed. Consider it the overall blueprint for any given cryptographic process. The algorithm is what's referenced when an encryption program alters your original data. The same algorithm must be used to reverse the encryption process. With symmetric-key systems, the private key used by the algorithm must be transmitted securely to the trusted parties before any transfers of encoded data can take place. This is one of the greatest challenges in the cryptography routine and will lead to our discussion of the two most common types of encryption and the algorithms they employ.

The length of the key it uses generally measures the strength of any given algorithm. A *key* is simply a string of code that tells the algorithm exactly how to scramble the data it's working with. When this same key is used in a reverse iteration of the original algorithm, the data is altered from *ciphertext* back into *plain text*. We've all heard the term, *128-bit*, used to measure the strength of Internet browser security. This number is simply a benchmark for what's called *strong encryption*. Any encryption process using less than a 128-bit key can't be called *strong*. Until January 2000, the U.S. government had an export ban on products using 128-bit encryption.

The number of *rounds* it goes through in its process also determines an algorithm's strength. Think of a round as a single stir in the mixing pot. The more rounds the algorithm goes through, the more stirred up the data gets. Different algorithms also encrypt data in *blocks* of varying sizes. These blocks, typically 64 to 128 bits, are just slices of the original data, broken into smaller, more manageable chunks. By dividing the data in this way, cipher strength can be increased while processing time is reduced. First, individual blocks are encrypted and then the resultant blocks of encrypted data are combined.

Although it's highly unlikely that you'll need to memorize the math employed by a particular algorithm, the exam will require your knowledge of the most commonly used ciphers. You should also be able to distinguish which category an algorithm belongs in—symmetric, asymmetric, hash, and so on. Let's take a look at the three basic categories of algorithms in use today.

 An algorithm provides a road map for any given cryptographic process. The key an algorithm uses gives it the information needed to apply its math in a unique way.

SYMMETRIC ALGORITHMS

As we briefly discussed in Chapter 3, *symmetric* (or *private key*) encryption uses a single key to encode and decode messages. This approach is fast and capable but the problem lies in exchanging the private key securely. Before any encrypted data can be exchanged, both parties must possess the same key. In some cases, keys are exchanged through means that are other than digital, such as snail mail or at a meeting. Believe it or not, people still have meetings, shake one another's hands and occasionally exchange private keys. Because of this "key" issue, symmetric encryption techniques are not fit for dealings such as Internet e-commerce. However, the private-key method is still widely used in environments where the benefit of its speed is vital and the key exchange dilemma is inconsequential. These situations include large data transmissions, recurring transfers of data between the same parties, smart card implementations, and local storage of encrypted information. Here are the algorithms native to the symmetric encryption method that you should know intimately.

 Symmetric encryption is faster than asymmetric but because the secret key must be exchanged, it is not suited for all transactions.

Advanced Encryption Standard (AES)

Regulated under The National Institute of Standards and Technology (NIST), the *Advanced Encryption Standard (AES)* is the Federal Information Processing Standard (FIPS) algorithm used by U.S. government organizations. NIST has recently selected the Rijndael algorithm as a replacement for the DES algorithm, which was the previous government standard.

Data Encryption Standard (DES)

DES, developed in 1975 by IBM and specified by the American National Standards Institute (ANSI) in 1981 as ANSI X3.92, is among the most popular symmetric-key systems available today. DES is known as a *64-bit block cipher*, which means it works with 64-bit plain text blocks and generates 64-bit ciphertext blocks as a result of its process. Using a 56-bit key, DES encrypts the 64-bit blocks of data using a 16-round algorithm. A 64-bit block is equal to 16 hexadecimal numbers. After the data has been sectioned into 64-bit chunks, each chunk is subsequently *permutated* (or systematically rearranged) to alter its bit order. Each 64-bit block is then divided into two, 32-bit blocks: a right and a left block.

The 56-bit key that DES uses is actually 64-bits long. However, every 8th bit in the key (bit numbers 8, 16, 24, 32, 40, 48, 56, and 64) is not used by the algorithm, but rather, is used for error detection. Now the 48-bit subkeys (16 in all) used in the 16 rounds of the algorithm are generated. This is accomplished by a process called *key scheduling*. The key scheduling routine produces subkeys that are intended to emulate randomly produced keys. At this time, the 56-bit key itself is split into two 28-bit parts. Each half is rotated, rejoined, and then expanded to 48 bits. These 48-bit subkeys are applied to each 32-bit block of data until all 16 rounds are complete. The data is now encrypted and the only way (theoretically) to decrypt it is to apply the original key with the same algorithm in reverse.

Note that this is a simplified explanation of the DES algorithm offered to give you a basic understanding of its inner workings. DES only uses a 56-bit key, technically classifying it as an all-around *weak* cipher. Another frailty of DES, and other block ciphers as well, is the weak-key factor. DES will occasionally generate *weak keys* resulting in regularities in its encryption. This unlikely but possible occurrence can provide a way to recover plain text within an encrypted message. A set of weak keys can have identical encryption and decryption characteristics, increasing the chances of the encryption being cracked. Although feasible, understand that this scenario is highly unlikely and the number of weak keys, compared to the number of possible keys, is more or less trivial.

So, why apply DES only once when you can apply it three times at triple the cost?

TRIPLE DES

Very simply put, *Triple DES* (also known as *3DES*) is just DES applied three times. The effective key length is a hefty 168-bits. *Voilà*, we now have strong encryption. It comes at a price, though. Triple DES slows down the process somewhat. However, it's still faster than other private-key methods available. As defined in ANSI X9.52, there are several possible modes of operation used by Triple DES. One method is to make all three keys unique. Another is to make keys 1 and 2 unique but key 3 is simply key 1 repeated again. Still another method is to use the same key three times. When this last method is used, the effective result is a Triple DES that's backward compatible with DES.

RIJNDAEL

Developed by two Belgian cryptographers, Vincent Rijmen and Joan Daemen, Rijndael (which sounds like *rain doll*) is an iterated block cipher that supports variable key and block lengths of 128, 192, or 256 bits. Its name is derived from equal parts of both developers' names. Rijndael has its roots in Square, a previous project of the same two cryptographers. Accepted in October 2000, Rijndael has become the new Advanced Encryption Standard (AES), replacing DES. This algorithm can be implemented to operate at speeds that are uncharacteristically high for a block cipher of its kind. It is also well suited for implementation on smart cards as it can be installed and run with a small amount of code using little RAM. It uses a four-step, parallel series of rounds making it stand up very well to cryptanalytic attacks.

Rijndael is also known for its ability to work proficiently in a wide range of hardware environments on various types of processors. It is also purported to work well over voice and ISDN lines, ATM networks, and satellite transmissions, but its developers state that as long as the processor in question can support the algorithm, it can be implemented in these and many other ways.

Rijndael was selected as the Advanced Encryption Standard (AES).

NOTE

BLOWFISH AND TWOFISH

Developed in 1993 by Bruce Schneier, founder of Counterpane Internet Security, *Blowfish* is a symmetric 64-bit block cipher employing variable length keys of 32 to 448-bits implemented in a 16-round process. It can be used as a drop-in substitute for the more time-consuming algorithms, IDEA and DES. Blowfish is unpatented, royalty-free, and requires no license to use.

Twofish, a 128-bit block cipher, is the follow-up project to Blowfish. It uses 128, 192, or 256-bit keys in a 16-round process. It is much faster than Blowfish and is also unpatented and free to use. The Twofish algorithm is efficient for use on smart cards and upon implementation, allows for performance trade-offs between speed, key scheduling, and memory usage.

SERPENT

A finalist in the AES competition, *Serpent* ended up with 59 votes in the final round, coming in second to Rijndael, which won with 86 votes. Serpent, designed by Ross Anderson, Eli Biham, and Lars Knudsen, is a 128-bit block cipher using 128, 192, or 256-bit keys in a 32-round process. While slower than Rijndael, (and indeed all of the other finalists, as well) Serpent is still faster than DES. It uses a relatively simple, straightforward process and its developers claim that although slower, it's more secure than Rijndael. I'll let you, future security professional, do your own cryptanalysis and come to your own conclusions.

SKIPJACK

Developed by the United States National Security Agency (NSA), the *Skipjack* algorithm is classified as *secret*. This means its details have not been released to the public for scrutiny. As this led to much debate on its level of security, the government invited a group of cryptographers to dissect the algorithm in private and they concluded in a report that it was indeed secure. The Skipjack algorithm processes 64-bit data chunks with an 80-bit key in 32 rounds of processing. It is the algorithm used in the Clipper chip. As a result of its secrecy, implementation of the algorithm is limited to government-authorized hardware manufacturers: it cannot be used in software.

INTERNATIONAL DATA ENCRYPTION ALGORITHM (IDEA)

Widely known for its use within PGP, IDEA is a 64-bit block cipher that uses a 128-bit key in an 8-round process. The algorithm effectively uses 52,

16-bit keys, which it generates from the original 128-bit key. The 128-bit key is first split into eight 16-bit keys. The original key bits are then shifted to make a new key. That key is then split into eight 16-bit keys. This process is reiterated until there are 52 keys.

It is said to be faster than DES but reports show that software implementations of IDEA operate at relatively equivalent speeds of those using DES. Hardware implementations of IDEA, however, prove to be somewhat quicker than DES.

MARS

MARS (Multiplication, Addition, Rotation, Substitution), IBM's submission to AES as a replacement for DES, is a shared-key block cipher that supports 128-bit data blocks and uses a variable key length of 128 to over 400-bits. It is purported to be faster and more secure than DES. The algorithm has a small footprint making it ideal for use in smart cards and other implementations where space is a concern.

RSA RC SERIES ALGORITHMS

Ron Rivest, lead developer of the asymmetric RSA algorithm, collaborated with some other great minds at RSA Laboratories to develop a series of private-key algorithms as well. The latest effort in the RC (*Ron's Code* or *Rivest's Cipher*) series—RC6—was one of the five finalists in the AES competition. RC6 addressed some problems that studies revealed within its predecessor, RC5, concerning some "interesting theoretical attacks." In line with the requirements of the AES, RC6 is a robust block cipher capable of handling 128-bit blocks of data. RC6 was designed so that the number of rounds, the size of the block it manipulates, and the size of the key are all adjustable. The key that RC6 uses has an upper limit of 2,040 bits. It is also well suited for hash functions, which we'll discuss later in the chapter.

Because the demands of the AES stated that the proposed algorithm work with 128-bit blocks, the RC5 design presented a complication. For RC5 to accomplish this, its design would have called for the use of two 64-bit *registers* (or chunks of memory), which would have been in contrast with the specified architecture for AES. The developers went back to the drawing board, and with RC6, presented a way to achieve this result using four 32-bit registers. Although RC6 was not selected for the AES, the algorithm is still used widely in many software applications.

RC2, an earlier algorithm in the series, is a variable key-size, 64-bit block cipher that can be used as a drop-in substitute for the slower DES algorithm.

RC4 is a speedy, variable key-size *stream cipher* found in products such as RSA SecurPC. A stream cipher encrypts data on the fly, making this algorithm work well in conjunction with SSL to encrypt data transferred between creators of secure Web sites and their customers.

A few other symmetric algorithms worth mentioning are the following:

- **GOST:** A 64-bit symmetric algorithm from the former Soviet Union. It employs a 256-bit key and can be used in software and hardware implementations.
- **Tiny Encryption Algorithm (TEA):** Designed by David Wheeler, TEA is a 128-bit cipher that uses a minimal amount of code to implement. TEA uses a large number of rounds opposed to a complex program.
- **CAST:** This is a *Feistel* cipher similar in style to DES. CAST comes is different strengths, namely 128 and 256. It is also implemented within PGP among other places.

There is an afterthought to the preceding discussion that should be addressed. You might be confused by the fact that two of the aforementioned private-key algorithms, IDEA and CAST, are used within PGP.

 PGP also employs Triple DES. Being aware that PGP is a public-key encryption method used for secure e-mail, you might have asked, "Why is PGP being discussed within the context of private-key algorithms?" The answer is that different types of algorithms can work together. The recent versions of PGP enlist the use of private-key algorithms for data transfers while using public-key algorithms such as *Diffie-Hellman* for the creation and exchange of keys.

 Cryptosystems can employ both symmetric and asymmetric algorithms, simultaneously.

ASYMMETRIC ALGORITHMS

Asymmetric (or *public-key*) encryption uses a public key and a private key. As we've discussed, the private key is never shared in this method, whereas the public key is freely distributed. Someone wishing to send a public-key encrypted e-mail, for instance, would encrypt it using the recipient's readily available public key. Public keys can be obtained through key-managing servers or from the recipients themselves. If the sender were to digitally sign the message, their private key would be used. Upon receiving the message, the recipient uses their private key, which is mathematically related to their public key, to decipher the message. If the e-mail message were digitally signed, the recipient would verify the signature with the sender's public key. Secure e-mail is but one implementation of the many asymmetric encryption methods in use. Nevertheless, we're here to talk about the algorithms that are fit to be classified as asymmetric, so here they are.

RSA

Named after its developers—Ron Rivest, Adi Shamir, and Leonard Adleman—the *RSA* asymmetric algorithm uses some elaborate math to deliver its brand of encryption. Originally developed in 1977, the algorithm was finally released into the public domain in September 2000. The RSA public-key cryptosystem is known for its implementation in the better part of all Internet e-commerce activities and is included in many popular software applications, such as Microsoft Internet Explorer and Netscape Navigator. This algorithm provides a means for encryption, authentication, and integrity verification, the latter two features being achieved by the use of Digital Signatures (DS).

This cipher bases its strength on the difficulty of factoring very large numbers. RSA uses huge *prime numbers* in its computations, resulting in distinctively strong encryption. You might remember that *prime numbers* are simply numbers divisible only by themselves and 1, such as 13. However, RSA uses much larger prime numbers exceeding100 digits. The RSA algorithm takes two of these large primes and multiplies them together. The resultant product of this computation, called the *modulus*, is the basis for the formula that RSA uses. In turn, the results of this formula are what RSA's public and private keys are comprised of.

Currently, there is no easy way to discover the private key if all one has is the public key. It is said to take an infeasible amount of time, with today's technology, to factor these large numbers out of a strong public key with a size of 2,048-bits, for instance. Of course, people are constantly developing new ways to make computers work faster and think harder, and if a discovery were made that enabled the quick factoring of these large numbers, RSA would become useless.

It is notable that in a challenge by RSA Security Inc., one of these numbers was indeed factored. On August 22, 1999, the factorization of a relatively small number (155 digits/512 bits) was accomplished. It took a group of researchers with almost 300 computers more than seven months to achieve. A 2,048-bit number would take an exponentially greater amount of time to crack and hasn't been done. If you're feeling optimistic, you can visit RSA Security's Factoring Challenge Web site at *http://www.rsasecurity.com/rsalabs/challenges/factoring/index.html* and give it a try.

Be aware that RSA is much slower than DES and other block ciphers. In software implementations, DES is roughly 100 times faster than RSA. It is also important to understand that key sizes for RSA need to be much greater than those of block ciphers such as DES in order to deliver similar strength. This contrast is due to the fact that in asymmetric methods a more detailed clue as to the value of the private key (in the form of the public key) is freely distributed for anyone to examine. For example, a 1,024-bit key size is recommended for use in a corporate environment, while a 2,048-bit key size should be reserved for only the most sensitive data transfers. These keys are 8 and 16 times larger, respectively, than the 128-bit *strong* key size of symmetric encryption. To make a long story short, the strength of an RSA key should not be compared to the strength of a DES or similar key based on its size alone. The two methods are like the proverbial apple and orange.

 RSA's strength is based on the complexity of factoring large numbers.

Diffie-Hellman (DH)

Developed by W. Diffie and M.E. Hellman while attending Stanford University in 1976, the *Diffie-Hellman* algorithm (or *key agreement protocol*) is the second most widely used asymmetric encryption algorithm next to RSA. Stanford held the patent for the DH algorithm until 1997 when it was released to the public domain.

Operating on a similar basis as RSA, DH provides a way to exchange private keys across an open connection such as the Internet without first exchanging any secret data. Instead of using the factoring problem that RSA does, DH provides security by way of the *discrete logarithm problem*. This mathematical dilemma is reportedly as hard to crack as RSA's technique and the formula that produces its values runs on CPUs at comparable speeds. Because of its strength, the discrete logarithm problem is implemented in other asymmetric cryptosystems such as ElGamal and DSS.

Elgamal

Elgamal is a DH-inspired public-key cryptosystem developed in 1985 by Dr. Taher Elgamal. Is it reported that there has never been a successful assault on this algorithm. As with DH, Elgamal gets its strength by means of the discrete logarithm problem.

 Asymmetric encryption uses two keys, a public key and a private key. Because the public key is out in the open, this method offers clues about the private key used to encrypt plain text. For this reason, asymmetric encryption requires larger keys than symmetric encryption to provide the same strength.

HASH ALGORITHMS

As we've learned, there are many uses for the hashing (or signature-only) process. These range from safely storing passwords to creating fingerprints of data that are used to apply digital signatures. The various hashing algorithms achieve their result through different processes. These processes are all irreversible. Two distinct features of hash algorithms are that input strings can be of any length but output strings have a fixed length. Hashing data is fundamentally different from encrypting it because a hash cannot yield the original data that created it. Nevertheless, hashing plays a vital role in cryptography. Essentially, to produce a *hash* (a.k.a. *message digest, fingerprint, md5sum*), a formula is applied to a string of text. The formula returns a smaller value called the *hash*. This condensed representation of the original data is the unique result of a specific formula applied to a specific string of text. The idea is that no other string of text (hashed by the same algorithm) will produce the same hash value. The hashing process makes it possible to do the following:

- **Store passwords ambiguously:** When a hash value of a password is stored on a system, the risk of passwords being exposed from an attack is diminished. A password's hash is generated whenever a user logs in. That hash is compared to the one that's stored on the system and access is granted or denied accordingly.
- **Verify the integrity of transferred files:** When you download a file from a trusted source, that source might also post a hash value of the file for viewing. Upon downloading the file, you can apply the same algorithm used to produce the original hash and compare the two values in order to verify that the file has transferred without corruption or alteration.
- **Create hash tables (or indexes):** A list of names, for example, can be converted to hashes, stored in a database, and made available for

query. Upon entering a name to look up, the system makes a hash of the name, finds its match in the database, and sends the user directly to that entry. This provides a search method that outperforms a standard, database-wide comparison search of all the entries.

■ **Verify digitally signed electronic documents:** The creation of a hash is an integral part of the DS process. Wrapped up within a DS is a hash of the original message, which enables a recipient to verify the signature and the message contents.

Let's examine a few of the algorithms that possess this unique ability.

Secure Hash Algorithm (SHA)

The *SHA* algorithm was developed originally by NIST in 1993 but technical revisions in 1995 led to the release of *SHA-1*, which is the version used today. Specified in Federal Information Processing Standards (FIPS) 180-1, SHA-1 is the U.S. government standard hash algorithm. The government also requires the use of SHA-1 with DSA.

The SHA-1 algorithm creates a 160-bit message digest in five rounds of processing. It also uses a process known as *message padding* to force the size of the pre-hashed text string to be a multiple of 512. Although it's possible, SHA-1 does not typically use any shared secrets or keys to accomplish this fingerprinting. Instead, security is provided by the fact that it's computationally infeasible for two different strings of text to produce identical message digests. More importantly, it's infeasible that an **altered** string of text would produce the same fingerprint as its unaltered, original counterpart. This is why a hash value comparison is a trusted method for verification of data. So you see, the security of SHA-1 is *assumed* because of the mathematical improbabilities laid out by this algorithm. If an assumption of security gives you an uneasy feeling, great!

You must be paying attention. Although not required by the exam, comprehension of the math behind SHA-1 (and any of the other algorithms) is the only way to wash away your fears. It is recommended that as a future security professional, you gain a broader understanding of these mathematical concepts. However, it's okay to wait until you pass the test to take that step.

 SHA-1 is the U.S. government standard hash algorithm.

MD5

MD5 is the latest in the MD series of hash algorithms developed by Ron Rivest. Introduced in 1991, MD5 addressed some security weaknesses found in its predecessor, MD4. Using a four-round process, this algorithm produces a 128-bit message digest, making it faster but less secure than SHA-1. MD5's formula also includes a step that pads the message length before hashing occurs. MD5's source code is freely available on the Internet and it can be put to use in a wide array of software and hardware implementations.

 Hash encryption produces a fixed-length string that represents data in a unique way and does not actually contain the encrypted data.

STANDARDS AND PROTOCOLS

Unless achieving the Security+ certification is but one step in your ultimate goal of becoming a security related software developer, you will most likely be using the readily available tools that have become standard. The evolution of cryptography began long before the computer came around but the digital family of techniques is by far the most effective to date. This family has some old-timers that are still employed as well as the many newcomers that are constantly changing the way we share our secrets. There are numerous security related *protocols* that are an integral part of this family.

A *protocol* contains a set of instructions for transmitting data from point A to point B. Applications and hardware devices are developed to take advantage of the benefits that a given protocol offers. This is as true in the field of security as it is in the online gaming world where the need for speed determines which protocols are used. In cryptography, speed is considered but ultimately, the strength of the cipher is what counts most. Some protocols are faster than others. Some are free to use while some are not. Some have superb encryption capabilities while others are merely acceptable.

Let's take a closer look at the security-flavored, protocol-enriched alphabet soup that you'll likely get a mouthful of while sitting at the exam.

SECURE HYPERTEXT TRANSFER PROTOCOLS (SECURE HTTP)

SHTTP (Secure HTTP) is a message-oriented communications protocol developed for use alongside HTTP. Its developers sought to offer a protocol that would work naturally with HTTP applications, emulating HTTP's methods and language rules.

SHTTP can support many different key management schemes, trust models, algorithms, and encapsulation techniques. Its flexibility makes it well suited for client/server-based Web applications such as online purchases that require sensitive data to be entered into forms. Because SHTTP supports end-to-end secure transmissions, sensitive data is encrypted from the moment it leaves a user's screen. This distinguishes it from older HTTP security schemes where the client had to first send data in plain text and be denied access before any encryption was in effect. Multiple cryptographic formatting techniques can be integrated into SHTTP clients and servers such as PKCS-7 and MOSS.

SSL (SECURE SOCKETS LAYER)

The *Secure Sockets Layer* protocol, developed by Netscape Communications Corporation, is the industry-standard technique for securing browser-based transmissions. It's what's being implemented behind the scenes when you visit a site whose URL begins with *https://*. Because it's built into all major Web servers and browsers alike, SSL stands ready for operation once a digital certificate is installed on the server.

Utilizing public-key cryptography, SSL supports 128-bit session keys providing strong encryption for events such as Web-based purchases. When you request an SSL secured URL, the Web server responds by sending out its digital certificate, thus authenticating itself. Your browser generates a unique session key, encrypted with the server's public key, which encodes all communications with the Web server. This method ensures that the Web server you are communicating with is the only machine that will be able to decipher your transmissions. Once a SSL session has been established, you might see the telltale padlock appear in your browser's window, indicating that a secure session has been established.

 The use of SSL requires that a certificate be installed on the server that authenticates it.

SECURE/MULTIPURPOSE INTERNET MAIL EXTENSIONS (S/MIME)

Modeled after the Internet MIME standard, *S/MIME* provides the ability to transfer secure MIME data. Originally conceived by RSA Data Security, Inc., S/MIME creates messages in the PKCS #7 data format and follows the

X.509v3 format for digital certificates. S/MIME is simply a version of MIME that adds public-key encryption for secure e-mail transmissions. Almost all corporate e-mail software developers have embraced S/MIME and provided products that support it. S/MIME presents us with a secure e-mail standard that provides authentication, encryption, message integrity, and proof of origin using Digital Signatures. Although Internet messaging is where S/MIME is most widely used, it can also be put to use by any method that supports MIME data—for example, HTTP transfers.

SECURE SHELL (SSH)

SSH is a standard protocol for securing remote connections over the Internet. SSH is used widely for the encryption of all data transferred between two configured ports. SSH runs at the Application layer of the TCP/IP stack and is broken into three major parts.

First, the Transport layer protocol (SSH-TRANS) provides server authentication and can also provide data compression. Secondly, there is the user authentication protocol (SSH-USERAUTH), which authenticates the client to the server. Finally, the connection protocol (SSH-CONN) handles the tunneling of the data stream in an efficient manner.

SSH can support a variety of public-key and symmetric algorithms including DSA, RSA, DES, Blowfish, Twofish, and more. SSH is used also in remote management solutions where an administrator needs a secure path to a server across the Internet. It is even touted as a total replacement for the FTP and Telnet protocols as it provides all of their functionality while enabling strong encryption.

DOMAIN NAME SYSTEM SECURITY (DNSSEC)

When the DNS system was conceived, security was not part of the plan. Because almost every Web-based transmission relies on this vital hierarchical database of sever names and IP addresses, it became clear that a method of securing this vulnerable system was needed. *DNSSEC* addresses this concern by using public-key technology to generate a Digital Signature on the zone data contained within a DNS server. The security concern is that if DNS data were altered by an unauthorized person, a user might be led to a masquerading Web site upon entering a trusted URL. This could result in the inadvertent release of information (such as a credit card number), or a rouge competitor actually stealing customers.

To ensure that DNS data is authentic, DNSSEC (which is actually multiple extensions to the DNS protocol) offers a way for administrators to create a hash value of the DNS data and then encrypt it using a private key. Web surfers essentially present the public key of the key pair used in order to decrypt the hash. If there is a match, then the data is considered good. The added amount of data required by the DNSSEC enabled DNS server could occupy almost 10 times as much space, making the price tag potentially high for secure DNS.

Additional security-related protocols are discussed in Chapters 2, 3, and 4. A few that should be reviewed include PAP, CHAP, TACACS, RADIUS, IPSec, and Kerberos.

PUBLIC KEY INFRASTRUCTURE (PKI)

The *public key infrastructure* (PKI) standard provides a method of transmitting data and exchanging money securely across insecure networks. A PKI is a network of services, procedures, and encryption software that work in conjunction to create a secure loop. At the heart of a PKI system is the *Certification Authority* (CA). The CA is a trusted third party that centralizes the issuance, management, renewal, and revocation of Digital Certificates. PKI employs a complex, multifaceted set of standards, procedures, and policies that are outlined in the following sections.

CERTIFICATES

A digital certificate contains a few vital pieces of information that vary depending on who or what it is certifying. These elements typically follow a format defined by the X.509 standard. A common certificate is comprised of the following:

- **Identifying information:** This includes the name and other identifiers of the entity in question. This could be an individual's full name, e-mail address, photo, and so on. For a server certificate, this section might contain the server's host name, IP address, and other information.
- **The public key:** This is simply the public-key data of the certificate holder. These days, the public key will most likely be an RSA key. However, some certificates might use another form of asymmetric key.
- **The signature of the Certification Authority (CA):** This element of the certificate is what validates the whole package. When the CA

applies their Digital Signature (DS) to the certificate, they are in effect making the statement that the information it contains is true. Because a CA is putting their name on the line here, you can be sure that they will confirm this information themselves before signing it. This can be accomplished by any number of electronic and/or paper-based authentication methods.

- When a DS is checked and verified by an individual or their software package, it can be assumed that all of the other data contained in the certificate is factual. Note that a private server within an intranet can be used to issue certificates exclusively to company employees or guests. In this private setting, authentication methods can be customized to fit the needs of the organization and would likely be less severe than those used in a public setting.

 A CA is the trusted body that issues and manages certificates.

Certificate Types

There are two basic types of certificates–*server* and *personal*. Each is assembled in a similar manner while serving different purposes.

Server Certificates

A *server certificate* can be put to use for many reasons. For example, a company that issues software updates online might have a certificate on the Internet server that hosts the updates. When a user requests an update, their software contacts the server to see if one is available. If there is an update and the user selects a download, the software can initiate the validation process. The server hands over the certificate, the software checks the DS, and if everything looks good, the update is applied without the user being bothered by the whole process. If not, the user would be warned of the bad certificate for several reasons such as an expired certificate. This prevents corrupted or altered versions of the software from making their way into a user's system.

If a company wishes to conduct secure transactions on their Web site, they would likely employ SSL. To put SSL to use, the server needs a certificate—preferably one from a well-known, trusted source. To obtain an SSL Web server certificate from a trusted entity for use in e-commerce, a company must first prove several aspects of their identity including their right to use the company name, their right to use the domain name, proof of phone numbers, and so on. An issuing CA might ask for bank statements, phone records, copies of passports or drivers licenses, fictitious name records, tax records, articles of incorporation, and so on. Simply put, you can't just ask for a certificate and then open shop—you will need to convince the CA of your identity and your legitimacy.

There is also a fee involved when obtaining such a certificate. However, the CA is not in business simply to sell certificates—they are after a much more lofty goal. The CA seeks to provide the Internet community with a quick and reliable way to help decide whether or not to do business with a particular company. The trusted certificate in this scenario ensures potential customers that data transfers between them and a company's server will not be intercepted and that the company has proven itself to be a stand-up operation.

Personal Certificates

A *personal certificate*, which can be obtained at no cost from some CAs, enables users to sign and encrypt transfers such as e-mail messages. The steps in this process are as follows—note that they employ the use of public-key cryptography.

Two users, Bryan and Drew, want to share some classified data over the Internet. Bryan has chosen the PKI method for the secure transmission and has obtained a digital certificate that proves his identity. To enable these users to exchange encrypted messages, Drew must get his own digital certificate. Drew then contacts a CA, proves he is indeed Drew, and is issued a certificate. Within the certificate is information about who Drew is, a copy of his public key, the CA's digital signature and an expiration date. Drew will also receive the other half of the key pair—the private key—which he will keep to himself.

Now that the CA has established Drew as a public key holder, Bryan can obtain Drew's public key (as easily as anyone else can) from a certificate management system. Bryan can compose his secret message and instruct his PKI software to use Drew's public key to encrypt the message. Bryan's software will also use his own private key to create a digital signature that will be attached to the message. The digital signature simply proves to Drew that Bryan did indeed compose the message.

Now the message is wrapped up in its digital envelope, signed, and sent to Drew. These steps ensure that only the intended recipient will be able to read the message and if its contents are altered in transit, the recipient will know about it. Drew receives the encrypted message and uses his private key to decrypt it. Because the message was encrypted with Drew's public key, only his private key can unlock it. Furthermore, using Bryan's public key, Drew can authenticate the digital signature contained in the message, proving that the message is really from Bryan.

PKI attempts to emulate the paper world by providing confidentiality, authentication, data integrity, and accountability to digital correspondents. Depending on the software used, these steps are nearly transparent. We might call the process translucent to the user, as there are a few steps that at first can confuse and/or vex a novice.

PKI has its share of critics because the process is not as easy as people would like it to be. Which CA can really be trusted? There are so many to choose from. On which digital key chain should all of these keys be kept? If keys are left on a work computer then messages can't be read at home. Also, keys are only as safe as a computer.

Despite its shortcomings, PKI provides a pretty solid method for secure communications. As far as future PKI systems are concerned, ease of use shouldn't come at the cost of a decreased level of security. However, until this process becomes more convenient, the Internet community might not embrace PKI as their standard for secure messaging.

In the preceding e-mail scenario, the users take an active part in exchanging keys and certificates. In other situations, your software program or browser does all the work and you might not even know that a certificate was involved. When dealing with certificates, whether behind-the-scenes or directly, it's important to know that the decision to trust an individual certificate is ultimately up to the certificate user. Luckily, there are some measures in place that assist in making this decision. Upon issuing a certificate, a CA is declaring that the public key (and other information) is bound to the individual or organization listed on the certificate. To supplement or limit that declaration, a CA can provide additional information in the form of standardized policies and statements.

Certificate Policies

As an amendment to the X.509 standard, a process of associating a *certificate policy (CP)* with a certificate was introduced. A CP is a collection of rules that states how an individual certificate pertains to a particular function and/or set of functions that share similar security demands. The CP exists to protect both the certificate user and issuer (the CA).

Upon examination of a CP, a certificate user can get the extra information needed to decide if the certificate is trustworthy. Likewise, upon association of a CP with a certificate, the CA is, among other things, protecting themselves from claims of loss or expense associated with certificate misuse. Represented within a certificate by a unique Object Identifier (OID), typically, a CP is detailed in additional documentation available from the CA. This information can be outlined in print and made available on the Internet. As described in RFC 2527, a certificate contains three extensions (or fields) that support the CP: *Certificate Policies extension, Policy Mappings extension,* and *Policy Constraints extension.*

Certificate Policies Extension

This extension, which has two variations, details the intended uses of the certificate. The first variation, called *noncritical,* lists the CPs that apply to the certificate in question. It is known as noncritical because the certificate is not limited to the uses outlined in the respective CPs. The other variation, called—you guessed it—*critical,* has a slightly different purpose. This type of certificate policies extension defines the restricted uses of the certificate as outlined in a particular CP. Also, this extension can include a pointer to a CA's *certification practice statement (CPS),* which we'll discuss shortly.

Policy Mappings Extension

This field is solely intended for use in CA certificates or certificates issued to one CA from another. It is found in certificates when two organizations have agreed to cross-certify one another's PKI. This extension defines the CPs that are predetermined to be interchangeable between the two entities. This provides an easy way of mapping CPs that exist in both organizations without the overhead of redefining rules within certificate-dependant applications.

Policy Constraints Extension

The constraints field can define one of two desired types of certification path behavior. A *certification path* is simply a hierarchical flow of trust between issuing CAs, similar to the trust between domains within a local or wide area network. The first choice inhibits policy mapping by disabling additional certificates in a certification path after a predetermined number of additional certificates have been issued. The number can be set to zero, if desired. This protects CAs from transitive trust security issues. If CA-1 trusts CA-2 and CA-2 trusts CA-3 but CA-1 does not wish to trust CA-3, CA-1 could disable policy mapping.

Conversely, the other option requires the enforcement of an explicit CP in a certification path after the issuance of a predetermined number of additional certificates. Instead of cutting ties a few steps down the path, this option simply enforces a CP once the CA in a trusted domain is effectively certifying outside that domain.

The CP exists to protect both the certificate user and the CA.

Certification Practice Statements

The *certification practice statement (CPS)* is simply documentation containing a CA's rules and regulations regarding certificate usage. It outlines the details of how a particular organization operates its network of trust.

A CPS is sometimes considered part of the agreement between a CA and a certificate user. This is a declaration that each CA customizes based on its own strengths and areas of specialization. Although it's recommended that any standards in use by a CA be discussed within a CPS, it has no set syntax or precise format. The CPS exists to provide information to potential certificate users or prospective partnering CAs. This information can assist individual users with any questions of trust they might have. It also provides an easy way for potential partnering CAs to determine if the technology employed by the CA in question is well suited for interoperability with their systems.

VeriSign, Inc., a leader in the field of electronic trust services, has a CPS posted on their Web site that outlines their practices and procedures. If you happen to be near a terminal, the latest version can be viewed at *https://www.verisign.com/repository/CPS/.*

 A CPS will include instructions for certificate use and a CA's provisions on how it suspends, revokes, or destroys certificates.

Revocation

When any certificate is issued, it is given an expiration date. This can be any amount of time but typically you'll find that the average certificate lasts for a year or so. When the certificate expires, a new one must be issued in its place. The main reason for this should be obvious; certificates are like driver's licenses or credit cards. It's just a bad idea to have valid certificates hanging around out in cyberspace for decades. Also, the more a particular public key is used, the more clues a hacker can gather to determine its private component.

There are occasions, however, when a certificate must be *revoked* before its actual expiration date. This might become necessary if the private component of a key pair is lost or discovered by a third party. Another reason could simply be that some of the supplementary information within a certificate has become outdated, necessitating a reissuance.

Most CAs also facilitate the *suspension* of certificates. Suspension might occur if the certificate in question is associated with a temporary problem and after the problem is fixed, the certificate will be listed as valid once again. There are several methods that enable software to determine if a certificate has been revoked or suspended but we'll discuss the two most common, CRL and OCSP.

Certificate Revocation List (CRL)
The *CRL* is the most widely used method for certificate status lookup. This list, which is made available on the Internet, can be queried via LDAP or transferred via HTTP or other means. When any certificate is issued, it is also assigned a unique serial number. The CRL contains a searchable database of all revoked certificates from a particular CA. When a program queries this list, it submits the serial number and obtains the certificate status in return. If a revoked, expired, or suspended status is returned, the application will reject the certificate. A software program might query a CRL only when needed or periodically download the entire list and check certificate status on the local machine.

Online Certificate Status Protocol (OCSP)
An *OCSP server* (or *responder*) is at the heart of this newer method for checking certificate status. One advantage of this protocol is that it can operate across TCP/IP networks in conjunction with the most common protocols. This enables an easier implementation of status lookup in client applications. Because OSCP supports the typical client/server method of data retrieval using TCP/IP standards, it's more compatible with applications and networks already in place. OCSP is touted as either a substitution or a supplementation to a CRL as most OCSP servers are getting their information from a CRL anyway. OSCP just packages the data in a form that's easier to retrieve and can also supply additional information not available in a CRL.

TRUST MODELS

As we learned in our discussion of certificate policies, it is common practice for CAs to cross-certify each other's PKI. This enables organizations to expand their base of trusted individuals and entities. Remember, a CA can still control the depth of trust regarding a specific certificate through the use of CPs. This type of cross-certification is a variation of the *hierarchical* trust model. A *trust model* is simply a series of rules that tells an application how to decide whether or not a certificate is legitimate. For our purposes, we'll discuss the two basic categories of trust models: hierarchical and Web-of-trust.

Hierarchical

This trust model (also called the *CA model*) is the basis for most certification systems in use. It is now the traditional model in use by large-scale certification authorities. In this model, certificate users place their trust in the CA instead of trying to come to their own decisions regarding the authenticity of a certificate. Once you feel that you can trust a certain CA, you are essentially agreeing to trust every other certificate the CA vouches for.

Hierarchical trust places the CA at the top level and the trust flows all the way down to the end user. This makes it easier for users because the burden is out of their hands. It's important to note that if a CA that you trust is cross-certifying another CA's PKI, your system will accept automatically the certificates of that CA as well. In high-risk situations, this might be undesirable. Comprehensive knowledge of a specific CA's practices is your only protection against unwittingly accepting the certificates of strangers. Know your CA well!

Web-of-Trust

This model is best known for its implementation in PGP. In the *Web-of-Trust* or *keyring* trust model, there is no centralized organization making the decisions. The users decide whom to trust based on personal knowledge or on the opinions of other users that they already trust. For instance, if someone you know personally hands you their public key, it's safe to tell your software that the key is trustworthy. This is accomplished by signing the key. Later on, a user who receives your public key can determine the keys you've signed. If they decide to trust you and sign your key, they are in turn making the decision to trust all whom you trust. This is how the Web-of-Trust grows.

PGP servers maintain a database of keys and the signatures that have been added to them. This method works great for small groups but corporate or government environments usually require a centralized system. In fact, the X.509 certificate standard used by CAs does not by design allow this type of trust model. The main problem with a Web-of-Trust is thecareless or malicious user who signs bad keys. If just one person in the Web-of-Trust turns evil, the whole group can be affected.

The trust model that an organization chooses depends on their particular needs. There are also new and crafty ways that enable the combination of both of the aforementioned models. These implementations grow with the research of the cryptographic community and are only limited by the imagination of software developers.

KEY MANAGEMENT/CERTIFICATE LIFE CYCLE

Because the algorithms used to generate the cryptographic components of a certificate are so strong, direct attacks on keys are fundamentally infeasible. Therefore, hackers will most likely aim their attacks on the key management systems that store and protect the keys. For this reason, it is imperative that a key management system be secure. If the security of a key server is compromised, a malicious user could impersonate someone else given that stolen keys can be used to forge certificates. A hacker could also alter keys in such a way that they could intercept "secure" transmissions.

Key management refers to the secure systems that generate, store, distribute, and provide information about keys. Although this data must be vigorously protected, at the same time, it must be available to established users. Key management techniques are designed to fulfill these needs simultaneously. The elements of key management include the creation, storage, protection, distribution, status checking, escrow, revocation, suspension, renewal, and destruction of keys and the certificates that contain them. While most of these functions are secured by a centralized system, private-key protection can ultimately be the responsibility of the user. There are many applications and protocols designed to provide these functions but few standards have been agreed upon. There are two broad categories of systems that aim to serve these purposes: *centralized* and *decentralized*.

Similar to the hierarchical trust scenario, *centralized* key-management systems refer to those that place the authority in a top-level entity. Again, this is the standard model used by large-scale CAs, corporations, and government agencies. It gives an administrator system-wide control regarding the many facets of key management.

Conversely, *decentralized* (or *manual*) key management allows the configuration of smaller-scale systems to be controlled by the user. Again, we turn to PGP for an example of this method. If a user were to misplace their PGP private key, they could revoke the key themselves and then notify the server of their actions. The server would then list that particular key as expired. In a centralized system, one might instead contact the administrator to do this for them. Also, decentralized systems do not provide all the functionality of centralized systems.

There are numerous vendors that package these management techniques into applications as part of their proprietary PKI solutions. There has been much heated debate on issues regarding key escrow and the introduction of *trusted third party* (TTP) entities that allow government access to private keys. Again, the reasons for choosing one management technique over another depend on the needs of the organization.

Let's look at some methods of key management and the different phases a typical key goes through in its life cycle.

STORAGE

Key *storage* is a multifaceted term that is relevant to the storage systems of private companies, CAs, TTPs, and end users, alike. It can refer to systems that store and provide access to public keys, such as searchable Lightweight Directory Access Protocol (LDAP) based directories. Because the public keys stored in a LDAP directory are indeed public knowledge, the confidentiality of these systems is not a major concern. However, attacks on these systems can affect their availability and integrity. If a hacker takes down one of these servers, a PKI infrastructure that relies on it can be interrupted. When the directory is unavailable, a PKI application might be configured to use a backup authentication method that's not as secure. A well-planned attack could be aimed at exposing the weaknesses of this backup system. Furthermore, an attacker could alter the contents of public keys. In this scenario, a number of complications could be introduced into the PKI routine.

Key storage also refers to centralized systems that store the *private* component of a key pair in case one is lost or needs to be accessed by an authority such as your boss or a law enforcement agency. We'll discuss these key recovery and escrow mechanisms later in this section.

Finally, key storage also refers to the methods of private-key storage used by the individual key holder.

As far as storing your own private key goes, there are some considerations to take into account. Generally, you will store your key in either software- or hardware-based mechanisms. Because a private key should never be stored in plain text form, the most common software based methods involve encrypting the private key with a password. This can be unsafe if anyone has access to your computer because passwords can be guessed.

Some software will allow protection of a key with an extended *pass-phrase*. A pass-phrase is like a password but it encourages a user to enter a much longer (and therefore, much stronger) string of characters. Pass-phrases can allow for spaces, so a whole sentence (complete with caps, numbers, and symbols) can be used as an entry. However, if you need mobile encryption/decryption capabilities, storing the key on a hard drive won't do. One solution is to carry the key on a floppy or CD-ROM. The security concerns regarding this practice should be obvious. Another method is to carry the key within a smart card. Smart cards can offer the added benefit of on-board processing, enabling encryption/decryption without exposing the private key.

An additional component of key storage that should be noted is the regular backup and archival of keys. As with any other type of sensitive data, a solid backup system is a must.

ESCROW

Key *escrow* is a term for the systems that provide a means of data recovery when the key used to encrypt the data in question is unavailable. These systems are comprised of a depository of private keys (usually with a TTP or key escrow agent) that can be accessed only by authorized individuals, such as our big brother. The term *key recovery* is also applied to the systems that provide this kind of unfettered access. Normally, an administrator will generate key pairs, copy the private component, and then provide the end user with the pair. The private key would then be included in the escrowed database. Alternatively, end users can generate key pairs and then securely hand over the private key for inclusion in the database.

Some key management systems incorporate what known as *m of n control* to facilitate key recovery. Using this method, a separate key is generated that has the unique ability to recover other private keys in the system. This key is then actually split among many administrators in the organization. The idea is that *m* out of a total *n* of these individuals must decide collectively that key recovery is necessary. This prohibits abuse of the recovery process by any one individual.

Although it has its opponents, key escrow and recovery have become necessary evils as the use of encryption continues to rise. For instance, imagine that your lawyer has encrypted some data on a personal computer that will prove your innocence in an upcoming court appearance. Furthermore, imagine that the private key used to encrypt the data is on a disk in the lawyer's pocket. If this person is involved in a fatal accident and the disk is destroyed, who has access to the data? The answer is, nobody. If the destroyed disk held the only copy of the key, you're in trouble. However, if the lawyer (or the firm that employed the lawyer) had put the key in escrow, there would be a way for you to recover the vital information locked in cipher-space.

 Key escrow provides a way for authorities to obtain private keys. In this scenario, there might be a stipulation that the key be destroyed after it's been used.

EXPIRATION

As we discussed earlier, it's good practice to impose time limits on the life of a certificate. Typically, this is done to protect against cryptanalysis of the key in question. An additional reason for this time limit is to reduce the effects of a compromised key. If an attacker obtains a private key and stays quiet about it, the user might never find out that the key has been stolen. Instead, the thief could simply collect data in transit, decrypt it, and add it to their own database over time. Enforced periodic key changes, just like enforced password changes, keep this risk to a minimum. When checking certificate status, a returned state of *expired* will cause an application to reject the certificate. Similarly, a state of *revoked* or *suspended* causes the same rejection.

RENEWAL

It's recommended that certificates be *renewed* before they expire. When a certificate expires, a new one can be issued in its place using the same keys or using brand new ones, depending on the situation. Sometimes, the only necessary change that a new certificate warrants is an update to one of its attributes. In this case, a new certificate could be issued using the original keys that the expired version contained.

DESTRUCTION

When certificates expire or get revoked, it's common practice for CAs to *destroy* the certificates and any keys associated with them. This is usually accomplished by overwriting the key data. One common method of destruction is called *zeroization*, which simply means that all sensitive data is written over with zeros. It can also be specified that copies of private keys that reside in memory be destroyed as well.

KEY USAGE

The X.509 certificate standard includes an extension for specifying *key usage*. This extension simply specifies the intended purpose of the key in a certificate. A certificate can have more than one entry in the key usage extension. In addition, these entries can be marked critical or noncritical. If marked critical, then the certificate must only be used for the one of the designated purposes. When an application verifies a certificate, it checks this extension and compares the requested use of the certificate with the specified use or uses. If there is a discrepancy, the application will not accept the certificate but the key usage extension might have information on the whereabouts of a certificate (owned by the same entity) that fits the bill.

Here are a few examples of intended key usage:

- **KEY_CERT_SIGN:** States that the key is to be used for signing certificates. Only applicable to CA certificates.
- **DIGITAL_SIGNATURE:** States that the key is to be used for creating digital signatures.
- **DATA_ENCIPHERMENT:** States that the key can be used to encrypt common data.
- **CRL_SIGN:** States that the key is to be used to sign certificate revocation lists.

Because the key usage extension is comprised of preset values, there also exists a customizable, extended key usage field. This field can specify any organization-specific intended uses for the certificate that supplement or replace the purposes noted in the key usage extension.

 If a key usage extension is flagged *critical*, then the certificate must be used for the specified purpose.

Multiple Key Pairs

If an entity does have multiple key pairs used for different reasons—such as one for signing and one for encrypting—it will have multiple certificates as well. This presents another challenge for key-management systems regarding the creation, distribution, revocation, and all-around management of multiple certificates owned by a single entity. One reason this is done is if an employee quits their job, the company can maintain the encryption key in order to read the employee's encrypted documents. The signing key, however, would be revoked to prevent any prospective ill-intended document signing. Multiple key pairs can also have different strengths, strong keys for encrypting sensitive data, and weaker keys for encrypting more commonplace data. This enables faster encryption of the less sensitive stuff. Another reason for key multiplication is to provide a backup if the key is lost. Still another reason is that a spare key might be needed for entrustment with a key escrow agent. Whatever the reason, someone needs to keep track of all these keys. The complex applications that tackle the task of key management aim to simplify the process.

TEST TIPS

Here are the Test Tips for the basics of cryptography chapter. Read them at least twice. A good understanding of algorithms, keys, DSes, CAs, and certificates is necessary for certification success. Furthermore, it's likely that you'll be asked some form of the question, "Which keys do what?," regarding the transfer of encrypted messages. You might also be asked about the key size used by a particular algorithm or how many rounds an algorithm goes through.

Here you go:

- √ *Private-key* (or *symmetric*) encryption uses a single, secret key. The biggest problem with symmetric encryption is that both parties must possess the secret key before encryption can ensue. Symmetric encryption is faster than asymmetric.
- √ *Public-key* (or *asymmetric*) uses a combination of a public key and a private key. Asymmetric encryption requires larger keys than symmetric in order to provide the same level of strength.
- √ An *algorithm* provides a blueprint for encryption methods. Know the difference between symmetric and asymmetric algorithms.
- √ The more rounds the algorithm goes through, the more encrypted the data gets.
- √ A 128-bit symmetric key = *strong encryption.*
- √ The *Rijndael symmetric algorithm* was selected as the Advanced Encryption Standard (AES), replacing DES.
- √ The *RSA asymmetric algorithm* bases its strength on the difficulty of factoring prime numbers out of very large sums.

√ *Hash algorithms* create a fingerprint or message digest. They are only used for comparative functions. The data that produces a hash cannot be recovered from the hash.

√ *SHA-1* is the U.S. government standard hash algorithm.

√ If you encrypt data for a secure e-mail, the recipient's public key is used for encryption. When the recipient decrypts the message, only their private key is required. Read this carefully. If you encrypt **and** sign data for a secure e-mail, signing it requires the use of your private key and encrypting it requires the use of the recipient's public key. When the recipient decrypts the message, they use their private key but to verify the signature, they need your public key. There is a difference between just encrypting a message and both encrypting **and** signing one. Know the difference; you will probably be asked about it.

√ *Confidentiality* is the process of keeping data hidden from unauthorized persons. Confidential = secret. Encryption provides confidentiality.

√ Guarding the integrity of data means keeping it from being altered while in transit or while being stored. Integrity = unaltered. MD5 fingerprints ensure integrity.

√ When we provide non-repudiation, we are ensuring that people can't back out of electronic dealings. Digital Signatures provide non-repudiation.

√ *Digital Signatures* (DSes) make use of encryption algorithms and hash algorithms to provide data integrity and non-repudiation. DS is the electronic equivalent to the handwritten signature.

√ At the heart of a public key infrastructure is the *Certification Authority (CA)*. The CA is a trusted third party that centralizes the issuance, management, renewal, and revocation of digital certificates.

√ Certificates follow a standard called X.509 and include a public key, the DS of the CA, and information that pertains to the entity that the certificate was issued to. There are server and personal certificates.

√ *SSL servers* must have a digital certificate installed that proves the organization's identity.

√ The *certification practice statement (CPS)* is documentation that summarizes a CA's intentions regarding certificate usage.

√ A *certificate policy (CP)* is a collection of rules that states how an individual certificate pertains to a particular function. CPs protect both certificate users and issuers.

√ Certificates have expiration dates but can be revoked or suspended before their time is up.

√ The *Certificate Revocation List (CRL)* is the most widely used method for certificate status lookup. *Online Certificate Status Protocol (OCSP)* is another method.

√ In a *hierarchical* trust model, the CA makes the decisions about who to trust. CAs can cross-certify each other with this model. Understand the potential dangers associated with cross-certification.

√ In a Web-of-Trust trust model, the users decide whom to trust.

√ The elements of key management include the creation, storage, protection, distribution, status checking, escrow, revocation, suspension, renewal, and destruction of keys and the certificates that contain them.

√ In *m of n key recovery*, the idea is that *m* out of a total of *n* individuals must decide collectively that key recovery is necessary.

√ Keys are destroyed by *zeroization*, or writing them over with zeros.

CHAPTER SUMMARY

This chapter has provided a basic understanding of how encryption techniques are put to work. You should now be familiar with encryption, PKI techniques, algorithms, certificates, certification authorities, Digital Signatures, and the methods that individuals and organizations use to prove their identity. If these tools are misunderstood, putting them to use can be dangerous and counter-active. Only accurate knowledge of cryptographic tools can help you secure your data.

As with all certification exams, you might have read something here that you won't see on the test. Be aware that certain explanations are given to establish the framework for concepts and ideas. Thoughtful study of these essential concepts will make the difference between selecting the right answer or the wrong one.

Now, go through these review questions and when you're done, go through them again. If you miss an answer, learn why. Keep taking all of the practice tests in this book and on the CD-ROM until you score perfectly. The aim of this book is to zero in on the facts that are needed to pass the Security+ exam but don't forget there are many additional resources out there to help you understand these concepts. If you have Internet access, they are right at your fingertips.

REVIEW QUESTIONS

1. **Which of the following algorithms provides hashing?**

   ○ A. Rijndael.
   ○ B. MD5.
   ○ C. GOST.
   ○ D. DES.

   Correct answer = B

   The MD5 algorithm provides hashing. Rijndael, GOST, and DES are all symmetric encryption algorithms.

2. **What size message digest does the MD5 hash algorithm create?**

   ○ A. 256-bit.
   ○ B. 512-bit.
   ○ C. 128-bit.
   ○ D. 56-bit.

   Correct answer = C

   MD5 creates a 128-bit message digest. Answers A, B, and D are invalid.

3. **What size key does the DES algorithm use?**

   ○ A. 56-bit.
   ○ B. 128-bit.
   ○ C. 168-bit.
   ○ D. 2-bit.

   Correct answer = A

   DES uses a 56-bit key, making it inappropriate for the encryption of very sensitive data. Answers B, C, and D are invalid.

4. **Which of the following algorithms is symmetric?**

   ○ A. IDEA.
   ○ B. SHA-1.
   ○ C. RSA.
   ○ D. MD4.

   Correct answer = A

IDEA is a symmetric algorithm. SHA-1 is a hash algorithm. RSA is an asymmetric algorithm. MD4 hash algorithm is an earlier version of the MD5 hash algorithm.

5. **Which of the following is a weakness of symmetric encryption?**

   ○ A. Key distribution.
   ○ B. Speed.
   ○ C. Hashing.
   ○ D. Key scheduling.

   Correct answer = A

   The problem with symmetric encryption techniques lies within distributing the private keys. Speed is actually a strong point of symmetric encryption. Hashing produces message digests. Key scheduling is the process an algorithm uses to generate subkeys.

6. **Which of the following algorithms was selected as the Advanced Encryption Standard (AES)?**

   ○ A. RSA.
   ○ B. Skipjack.
   ○ C. Serpent.
   ○ D. Rijndael.

   Correct answer = D

   Rijndael was the final selection for the AES. RSA was a finalist in the AES competition. Skipjack is a secret algorithm developed by the NSA. Serpent was a finalist in the AES competition.

7. **The cryptographic strength of RSA is based on which mathematical problem?**

   ○ A. Discrete logarithm.
   ○ B. Factoring.
   ○ C. Long division.
   ○ D. Fuzzy math.

   Correct answer = B

   RSA's strength is based on the difficulty of factoring large numbers. The Diffie-Hellman algorithm uses the discrete logarithm problem. Answers C and D are invalid.

8. **Which element of a Digital Signature (DS) provides integrity verification?**

    ○ A. Time stamp.
    ○ B. Middle initial.
    ○ C. Message digest.
    ○ D. Password hash.

    Correct answer = C

    Comparison of the message digest in a DS (Digital Signature) provides integrity verification. Time stamps make a note of the time that a document was signed and provide a form of non-repudiation. Middle initial is a fictitious answer. A password hash is a secure method of storing passwords.

9. **Which of the following issues digital certificates?**

    ○ A. CA.
    ○ B. DS.
    ○ C. DNSSEC.
    ○ D. CRL.

    Correct answer = A

    A Certification Authority (CA) issues digital certificates. DS is a digital signature and is part of a digital certificate. DNSSEC provides an extension to DNS that enables secure DNS transfers. A Certificate Revocation List (CRL) is an online database of revoked certificates.

10. **Which trust model does the X.509 certificate standard support?**

    ○ A. Web-of-Trust.
    ○ B. Gradational.
    ○ C. Fragmentary.
    ○ D. Hierarchical.

    Correct answer = D

    The X.509 certificate standard supports the hierarchical trust model, which places an authority (a CA) in charge of trust decisions. Web-of-Trust lets users decide whom to trust and is not supported by X.509 certificates. Answers B and C are invalid.

11. **Which of the following is commonly used to secure HTTP transmissions?**

    ○ A. OCSP.

    ○ B. Zeroization.

    ○ C. DSL.

    ○ D. SSL.

    Correct answer = D

    With the assistance of digital certificates, SSL is commonly used to secure HTTP transmissions. OCSP is a protocol used to determine certificate status. Zeroization is a process used to destroy keys. DSL is a high-speed Internet solution.

12. **What technique lets law enforcement agencies access private keys?**

    ○ A. Escrow.

    ○ B. Revocation.

    ○ C. Expiration.

    ○ D. Cryptanalysis.

    Correct answer = A

    Key escrow provides a way for authorities to access private keys. Revocation is the process of revoking a key before it expires. Expiration happens when a key reaches the predetermined time that it's set to expire. Cryptanalysis is the art of determining the plain text version of an encrypted block by studying the ciphertext.

13. **The message digest in a digital signature (DS) is encrypted with the**
    _____.

    ○ A. Sender's public key.

    ○ B. Recipient's private key.

    ○ C. Sender's private key.

    ○ D. Recipient's public key.

    Correct answer = C

    When a DS is created, it's the sender's private key that's used for encryption. Answers A, B, and D are invalid.

14. **Where would you find the rules outlining how and when a CA destroys certificates?**

    ○ A. CRL.
    ○ B. CPS.
    ○ C. Certificate Policies extension.
    ○ D. OCSP.

    Correct answer = B

    A certification practice statement (CPS) contains documentation detailing a CA's rules and regulations. In a CPS, you would find (among other things) policies regarding certificate destruction. A certificate revocation list (CRL) is an online database of revoked certificates. The certificate policies extension is a field within a certificate that supports a certificate policy. OCSP is a protocol used to determine certificate status.

15. **Which process rearranges the bit order in a block of data?**

    ○ A. Permutation.
    ○ B. Incantation.
    ○ C. Multiplication.
    ○ D. Pontification.

    Correct answer = A

    Permutation refers to the process of bit order rearrangement. Incantation is a magic spell. Multiplication is an unrelated mathematical term. Pontification is merely the expression of an opinion.

16. **What does m of n control provide?**

    ○ A. Authentication of the key-sharing process.
    ○ B. Confidentiality of the key-signing process.
    ○ C. Integrity of the key-destruction process.
    ○ D. Restriction of the key-recovery process.

    Correct answer = D

    M of n control provides a means of restriction in the key recovery process. In a total group of "n" people, "m" of them must agree that the key recovery should proceed. The numbers of people in each group, m and n, can be customized to fit the organization. This method prevents one person from having control over the process and thus restricts a disgruntled individual who wishes to steal keys. Answers A, B, and C are invalid.

17. **PGP uses which type of algorithm?**

    ○ A. Symmetric.

    ○ B. Asymmetric.

    ○ C. Both symmetric and asymmetric.

    ○ D. Neither symmetric nor asymmetric.

    Correct answer = C

    PGP enlists the use of symmetric algorithms for data transfers and asymmetric algorithms such as Diffie-Hellman for the creation and exchange of keys. Answers A, B, and D are invalid.

18. **What type of temporary key does a Web browser create during SSL communications?**

    ○ A. Open key.

    ○ B. Session key.

    ○ C. Siesta key.

    ○ D. Weak key.

    Correct answer = B

    A browser creates a temporary session key that's discarded after the SSL session is over. Answers A and C are fictitious. A weak key is a phenomenon where an algorithm creates keys that cause regularities in its encryption, making the encryption easier to crack.

19. **Which of the following is a weakness of asymmetric encryption?**

    ○ A. Requires exchanging private keys over the Internet.

    ○ B. Recovery of encrypted data is impossible.

    ○ C. Requires larger keys than symmetric.

    ○ D. Requires larger hash values than symmetric.

    Correct answer = C

    Because the public keys exposed in asymmetric encryption methods provide clues about their private counterparts, asymmetric key lengths must be larger than symmetric key lengths to provide comparable strength. The need to exchange private keys is a weakness of symmetric encryption. Recovery of encrypted data is impossible with hash encryption. Answer D is invalid.

20. Which algorithm is known as a stream cipher?

○ A. RC4.

○ B. MD5.

○ C. MARS.

○ D. CAST.

Correct answer = A

RC4 is a stream cipher, which means that it encrypts on the fly. MD5 is a hash algorithm. MARS is a block cipher. CAST is a Feistel cipher.

REFERENCES

http://www.rsasecurity.com/rsalab5 is the RSA Laboratories online research facility sponsored by RSA Security Inc. This Web site offers information on cryptographic algorithms, encryption standards, and the latest security bulletins. It also contains general security-related facts and suggestions on where to learn more about cryptography.

http://www.verisign.com is the Web site for VeriSign—the leader in digital trust services. The *Verisign.com* Web site has a rich knowledge base and a comprehensive White Paper section with information on PKI and SSL technologies.

http://www.counterpane.com is the Counterpane Internet Security Web site. Counterpane maintains an online library full of security-related publications and other resources. They also send out a free monthly e-mail newsletter on cryptography and computer security.

http://www.esat.kuleuven.ac.be/~rijmen/rijndael/ is the home page of the Rijndael algorithm. Although no longer maintained, this site has links to more information about the Advanced Encryption Standard (AES) competition winner than you might ever want to read.

http://csrc.nist.gov/encryption/aes/ is the Web site maintained by The National Institute of Standards and Technology; it is the homepage of the AES. Among the cryptographic-related information, you'll find some interesting historical data outlining the progress of the AES competition.

http://www.ietf.org/ is an Internet Engineering Task Force Web Site that has information on how Internet standards come to be.

http://www.faqs.org/rfcs/ offers one-stop shopping for RFC's and also has links to many Internet-related standards organizations online.

http://www.pkilaw.com has a wealth of information on Public Key Infrastructure with a focus on the legal issues that have come to light amid this new technology.

CHAPTER 6

OPERATIONAL/ ORGANIZATIONAL SECURITY

IN THIS CHAPTER

- Physical Security
- Disaster Recovery
- Business Continuity
- Policies and Procedures
- Privilege Management
- Forensics
- Risk/Threat Identification
- Education
- Documentation
- Test Tips
- Chapter Summary
- Review Questions
- References

In this chapter, we will address all aspects of operational and organizational security as they are strictly specified in the CompTIA domain objectives. It is very likely that the Security+ exam will ask you a plethora of basic level questions regarding the topics that exist in this chapter. Although you have already made your way through the most difficult content chapters in this book, it is imperative that you understand the concepts and information contained in this chapter as well as the chapters that follow. Remember, all questions on the exam are weighed equally. You might answer a question regarding advanced cryptography concepts correctly. Do not follow that up by incorrectly answering a question regarding basic level backups or simple log storage retention.

It can be best said that this is a security concepts chapter. If you have studied for the CISSP (Certified Information Systems Security Professional), you will understand this meaning. It's has more to do with the important implementation of safeguards, education, policies, and procedures than actual software/hardware implementations or a "foot soldier's" point of view.

 Common sense will be your best ally when addressing operational/organizational questions on the Security+ exam. Do not try to "over think" the easy questions.

PHYSICAL SECURITY

Physical security has to do with the implementation of countermeasures and actions taken in order to protect resources and assets such as people, buildings, and information from theft and environmental destruction.

This is the Age of the Internet. It seems that most security discussions or "security hoopla" focuses on the protection of resources from cyber thieves and damaging programs on the Internet. All that is fine and dandy. However, direct internal access to a system or resource provides a major opportunity for someone with malicious intent to corrupt, steal, or damage your information easily. Physical security focuses on using physical barriers and devices in order to limit access to important resources for authorized personnel only.

Physical security also places emphasis on the protection of resources from damage that can be cause by such things as fire, electricity, water, earthquakes, and storms. Next, we will discuss the specific topics that CompTIA expects you to know regarding physical security.

ACCESS CONTROL

Throughout the first five chapters of this book, we have discussed controlling access to information such as files, folders, and system information. Physical security primarily places an emphasis on controlling access to sensitive physical areas.

One of the first things you need to consider when addressing access control is the development and implementation of a documented access control policy. Most businesses have employees that support external clients who visit the business. These employees have periodic maintenance and service personnel visit the main site and remote locations and have them receive deliveries. It is imperative that you have a procedure that identifies who has access to certain locations within a secured local or remote site. The procedure should contain instructions for such things as ID changes and verification if you use electronic pass systems.

The important need for an access control policy can be seen in the following example.

An electrical maintenance worker calls your building's security guard or you and requests access to one of your remote locations for "routine maintenance." You happen to have important server systems connected to your internal network at that location. What do you do? Does the security guard know what to do? The real-life answer to this question and for the real exam is based on what's in that little access control policy that you and your company have developed. In other words, you might have a digital pass system and ACL (Access Control List), a remote surveillance device, or biometric device that can be used to determine remotely whether the worker should be allowed or denied entry to the remote site.

 It is likely that the exam will target the need for in-place procedures and controls regarding the physical accessibility to remote locations.

Next, we will discuss the physical barriers that are implemented typically to control access; we will also revisit biometric access controls.

PHYSICAL BARRIERS

A collective system of physical separators (barriers), exit and entry controls, alarms, and physical intrusion detection devices should be implemented to truly detect, deter, and delay unauthorized access or malicious intrusion to a secured area. In today's world the combination of this type of separation and protection is not always feasible. Budget restraints, lack of personnel, and the idea that "nothing bad has ever happened to us before, so let's spend the money on something else," are usually the causes for weak physical security restraints.

The following are physical controls that are commonly implemented in order to place a *barrier* or form of protection between unauthorized personnel and sensitive locations or data:

- Guards
- Dogs
- Gates and fences
- Turnstiles
- Mantraps
- Biometric devices
- Magnetic identification cards
- Photo ID cards

BIOMETRICS REVISITED

As you might recall, *biometrics* were described in Chapter 2. For physical security study purposes and the fact that you will most likely encounter several questions relating to this topic on the real exam, here is a refresher.

Biometrics are human- or character-based authentication methods that allow or disallow real-time access to systems, resources, or physical locations based on physical characteristics or behaviors.

Biometric authentication devices usually require the person who desires access to a specific physical location or resource to be present at the time of authentication or identification. These devices also eliminate the need for remembering passwords or PINs. Biometric devices also eliminate the need for a physical pass, card, or token that can be lost easily or stolen.

ENVIRONMENT

The security threats from physical and environmental conditions can be limited by following proper standards, codes, and guidelines. It is important when planning for a new site as well as an existing site, that the location be as secure as possible and that the proper measures are taken to protect an environment from fire, water, and electricity as well as forms of possible sabotage. There are many precautions that can be taken to reduce possible damage that can occur as a result of unfavorable environmental conditions. CompTIA expects you to know the following subtopics and the precautions associated with them.

Location

Determining a new site location for your company can play a major role in the overall success and security of a business. It is of utmost importance that you choose a site located in an area with favorable conditions. Some conditions that you should look for are as follows:

- A site located in an area with a low crime rate.
- Multiple access paths into and out of the site location.
- A geographically stable site. For example, no fault lines, a low flood area, and no trash dumps.
- Away from airline, railway, and major construction paths.

Ultimately, the amount of money your company budgets will determine your overall success rate for a new site location project. Sometimes, it is more cost effective to build a new site from the ground up than to move into a previously occupied site. If you are provided the opportunity with the construction of a new site, here are some internal physical building and room specifications you will need to consider:

- Secured doors should be resistant to forcible entry and should unlock automatically in the case of an emergency.
- Ceilings should meet building fire codes. Avoid drop ceilings, if possible.
- Walls should meet building fire codes. Rooms containing highly sensitive machinery and data should have their own power circuits, air conditioning support, and have a higher fire rating than rooms with general access.
- Raised floors that meet fire ratings should be used for data and computer centers.
- Fluorescent lighting is less expensive than traditional lighting and conserves energy. Use fluorescent lighting if possible.
- All building air conditioning should be on separate dedicated power circuits.
- A backup generator with the capability of supporting building power should be located in a secure area away from general access.
- Certified professionals should install an internal sprinkler system.

Once all these physical characteristics are in place, responsibilities and procedures must be implemented in order to facilitate prompt and proper response to any emergency situation regarding these factors.

Shielding

Shielding important systems, media, electrical components, wiring, and secured areas from external and internal environmental threats should be a top priority in your security planning. It is very important that all critical media be stored in a secure safe location. Magnetic tapes, disks, CDs, tokens, and important documents should be stored in a fireproof vault of safe if possible.

Electrical power panels, generators, and larger redundant power systems such as UPS backup units should be in sealed areas located away from general access.

All doors providing access to secured areas should have the ability to close airtight. The doors should be fireproof and include a fire sensor that automatically unlocks the doors if fire is detected or a power failure occurs.

 For the exam, know that the doors, which have electronic auto-locking mechanisms, should be tested frequently for their ability to automatically unlock in case of fire or power failure.

Electrical and network cabling should not be exposed. If cables or wires are exposed currently, consider using a conduit or covering that meets wiring and building fire codes. Only allow certified electrical technicians to work with electric wires and circuitry in your building.

In Chapter 4, it was mentioned that fiber-optic cable is more secure than many other types of cabling media and is less susceptible to interference or *crosstalk*. If you can afford fiber-optic cabling, you should implement it for better network performance as well as its security benefits. Other types of cabling, such as CAT5, are not as secure and can be wiretapped easily. As far as coaxial cable, replace it if possible. It is highly susceptible to interference, it can be easily tapped, and its protective covering produces a poisonous gas if burned.

Fire Suppression

It is critical that you have the proper equipment and procedures in place to detect, prevent, suppress, and react to the physical security threat of fire. Fire and smoke detection alarm devices, sprinkler systems, and accessible hand held fire extinguishers are the best resources for fire detection and suppression.

There are various types and specifications associated with the devices just mentioned. The most important devices you need to be concerned with for the exam will be mentioned next. Keep in mind: the exam is likely to confront you with the proper suppression method that should be implemented in case of a fire. Only one of the choices will be valid, all others will contain inaccurate information. In simple terms, the exam will try to trick you here.

Fire Extinguishers

Handheld fire extinguishers should be placed in easy-to-reach locations throughout a facility. For the exam, be sure and know which type of fire extinguisher should be used for various types of fires.

You should be familiar with the four following types of handheld fire extinguishers:

- **APW (Air Pressurized Water):** An APW fire extinguisher is a large, silver handheld extinguisher that is filled with a combination of air and water. It should never be used to put out a chemical or electrical fire. This is an older type of extinguisher that is used primarily to take the heat element away from a fire.
- **Dry Chemical (ABC and BC):** These types of handheld extinguishers are very effective at putting out various types of fire. Dry Chemical extinguishers smother a fire with a phosphorous chemical that separates the oxygen and fuel within a fire. ABC type extinguishers can be used to put out chemical, electrical, or normal wood burning or paper fires. You can identify whether the extinguisher is an ABC or a BC extinguisher by the pictures and labels on the extinguisher itself. A word of caution: never use a BC extinguisher on a fire classified as A. *Class BC* fires are electrical and chemical only. *Class A* fires are normal paper/wood burning fires. Simply put, educate yourself on the types of extinguishers available at your facility. Chances are you have ABC type fire extinguishers proliferated around your building.
- **Carbon Dioxide (CO_2):** These types of extinguishers use carbon dioxide gas to remove or displace the oxygen in a burning fire. They can easily be identified by a hard black horn or spout used to spray the chemical. Carbon Dioxide handheld fire extinguishers are designed to put out BC type fires.
- **Halon-Halon extinguishers:** These are filled with a gas instead of a chemical powder. This gas is more effective at putting out ABC type fires than an ABC type extinguisher. Besides providing better fire suppression than the previously mentioned extinguisher types, a Halon extinguisher will not ruin whatever you have just saved from

fire destruction. The chemicals in an ABC type extinguisher will ruin electrical wires, computers, or anything else you use them on. Although Halon works well at putting out fires, Halon extinguishers are banned in many places. It has been scientifically proven that Halon gas depletes the ozone and is considered very dangerous to humans. A good substitute for Halon is *FM-200*. FM-200 is a widely accepted, chemically based, fire suppressor that extinguishes fire by cooling or removing the heat from the flames.

Sprinkler Systems

As stated earlier, your best weapons to combat fire within a building are fire extinguishers and sprinkler systems. As with the various types of hand held fire extinguishers, you should be familiar with the basic types of sprinklers and systems used to distribute water or Halon. This is an important issue because some types of systems will offer you the ability to turn a sprinkler system off if a fire is contained. Others will destroy your computers and other electronic devices as well as surrounding material. Sprinkler systems or pipe systems are classified into the following categories:

- **Wet pipes:** Water always remains in the pipes that lead to the sprinkler head or nozzle in this type of system. This can be a great system to have in the event of a fire that you personally are not able to contain. If the system detects fire, water is quickly sprayed over the area. However, there is little you can do to shut off this type of system if the fire has been sensed but you happen to personally contain it. You will have put out the fire and the system will spray everything anyway.
- **Dry pipes:** Water is held far back from the nozzle by a clapper valve with this closed sprinkler-head system. If the system detects fire, there remains significant time to shut down the system if you happen to put the fire out before water is needed.
- **Deluge:** A *deluge* system also uses a valve to hold back water in the pipe. However, it uses an open sprinkler nozzle or head to rapidly distribute water.
- **Preaction:** This is a closed-head system and is a combination of wet-pipe/dry-pipe technology. It uses a built-in alarm mechanism to warn before it distributes water. This type of sprinkler system is often found in and recommended for electronic data centers and computer rooms.

- **Gas Discharge system:** This is a system that does not use water. Instead, it uses halon, carbon dioxide, or FM-200.

PHYSICAL SECURITY BEST PRACTICES

In conclusion to our discussion on physical security, you should be aware of the following best practices:

- **Create a physical security policy:** You need to implement and document security practices and procedures that are designed to meet the needs of your specific location and business. For example, if your business is located in a hurricane zone, you need proper instructions that educate employees on what to do in case of a hurricane. Your policy should also include rules for "who is allowed where" and proper instructions for handling disaster prevention devices. Your policy should also include federal, state, and local regulations as they pertain to the use of specific emergency related equipment and rules.
- **Control access to all important areas containing valuable assets:** This includes server rooms, electronic rooms, and access to places where security devices, such as recording mechanisms and controls, are located or stored.

Also be aware of these best practices:

- Use human guards to monitor and review access to secured areas.
- Use combination keypads, magnetic proximity badges, and biometric devices to control access to secure locations.
- Audit and log all access controls.
- Test all security controls periodically to ensure they are in working order.
- Educate all valid and authorized employees on the need to know physical security practices, policies and proceedures.
- Use common sense.

OTHER SECURITY CONTROLS

There are several other types of operational security controls that you should be made aware of just in case the CompTIA exam decides to spring them on you. They are as follows:

- **Corrective controls:** These types of controls are typically implemented after a weakness has been discovered or a problem occurs. For example, hardening or patching a Web server after a breach has occurred or a new threat has been discovered. Another example: implementing more restrictive file level permissions after a breach has occured.
- **Detective controls:** These controls are used to track or identify security breaches. Examples of detective controls are implementing, auditing, and reviewing log files, and monitoring suspicious activity.
- **Preventive controls:** These types of controls are implemented as a means of preventing a security breach. Using NTFS file level permissions before confidential files have been viewed and implementing such things as antivirus protection and strong passwords are examples of preventive controls.
- **Recovery controls:** These controls include having the ability to restore or rebuild a system and/or network environment after a disaster or security related incident has occurred. Example: a good backup system.

DISASTER RECOVERY

Disaster recovery involves the processes that take place in order to put things back the way there were before a disaster occurred, or the ability to accomplish this task. In the technical business world, it is important that processes, hardware, software, and other necessary devices be implemented in order to achieve this goal if a disaster were to occur.

 The Security+ exam is likely to focus on your knowledge of the basic backup techniques and strategies that are normally implemented to backup and restore data. However, it is also very important to realize that a complete Disaster Recovery Plan (DRP) should include the ability to replace components such as circuit boards, servers, workstations, people (knowledgeable), and other critical company assets needed to continue business as usual if disaster strikes. In other words, backing up every possible piece of data in your enterprise will do you little good if you have nothing to restore the data to.

In order to understand what information you need to backup and what the possible threats are to your data and other critical company assets, you should always first carry out a risk evaluation. Once you have identified threats, risks, and other critical information, you can plan for backups, off-site storage, and possible alternative sites to carry out your critical business tasks.

BACKUPS

The creation of a copy of your critical files, programs, and configured operating systems is critical to the life of your business (and your job) in the event of hardware or software malfunction or other disaster.

Removable storage media such as floppy disks, ZIP disks, optical disks, tape cartridges, CDs, or removable hard drives can be used in combination with software to back up your data and store it off site in the case of a disaster or emergency. The type of storage media you use will depend on the amount of information you need to backup and the speed at which you need the backup to process.

There are many types of backup software packages available on the market that allow you to carry out various backup types and functions. Some of the most popular are ARCServe by Computer Associates and Backup Exec by Veritas. Most operating systems available today come with their own internal backup programs. For example, most versions of Microsoft's Windows come with Microsoft Backup. UNIX and Linux operating systems offer the tar (tape archive) command.

Backup Types

The backup type and plan you implement depends on the amount of data you have to backup and frequency at which the data changes. In other words, if the majority of your files, folders, and other data change consistently throughout the day, you might want to consider a full backup on a daily basis. However, if the majority of your data does not change quite so often, you might want to consider a one time weekly full backup and a daily incremental or differential backup.

GFS (Grandfather-Father-Son) is the name that is often used to describe a backup strategy that includes a daily, weekly, and monthly backup. This type of backup system considers the daily backup as the son; the weekly full backup is the father; and the last full backup of the month is considered the grandfather. This backup strategy is based on seven backups a week. A full backup is run once a week and an incremental or differential backup is run on the other days.

 Make sure you are familiar with GFS for this exam.

Whether you choose a daily incremental or a daily differential backup, it mainly depends on how much media you have available and how fast you want to be able to restore the data.

 Pay attention! A+, Network+, Security+, and countless other certification exams often focus on what type of backup method to use for the easiest or fastest restore in a given situation. **Do not miss the easy questions!** A GFS backup strategy using a daily differential backup provides the easiest and fastest restore! Be careful on the exam. Some questions might be provided to you with very minimal information and leave you wondering how they can ask such questions. Try to keep focus on the overall concept that CompTIA is targeting. The following site explains the GFS backup strategy in detail: *http://www.intel.com/support/storageexpress/6736.htm*.

Files and folders that reside on most popular operating systems have four basic attributes assigned to them:

- Read only
- Hidden
- System
- Archive

When a file or folder has been added, written to or changed, the archive bit changes. Backup system software uses the properties of this archive bit to determine whether or not the file or folder should be backed up based on this flagged archive bit. Keep this in mind as you study the following basic backup types that you will need to know for the exam:

- **Full:** This backup type includes all folders, directories, files, and programs that reside on a system or disk. A full backup resets all archive bits regardless of their status. Typically, it is not productive to run a full backup every business day unless you have an incredible amount of backup storage space or a very small business with few systems.
- **Incremental:** With this type of backup, only files that have changed or have been added since the last full or incremental backup are backed up. The archive bit is cleared or reset. If you have a large

amount of daily information to backup, you should consider a backup plan that includes one weekly full backup and a daily incremental backup. Remember, if you need to do a restore with this plan, you will need to use your weekly full backup tape and every incremental tape created since the last full for a proper restore. In simple terms, an incremental backup job will run faster than a differential. However, it will take longer to restore.

- **Differential:** With this type of backup, all files that have changed or have been added since the last full or incremental backup are backed up. This backup type does not clear the archive bit. A *differential* restore is simply restoring the last full backup tape followed by the most recent differential backup tape. In simple terms, a differential backup job will take longer than an incremental to backup. However, it will do a restore much faster than an incremental.

- **Copy:** A *copy* is simply a backup of files, folders, and directories that are copied to another location whether it is a network share, tape, or hard drive.

OFF-SITE STORAGE

It is critical that your disaster recovery plan include at least one off-site storage location for the critical data you backup on a daily, weekly, and monthly basis. If a disaster were to happen at your location, such as fire or other disaster, it might do you little good to have the only copy of your important data stored in that same site. Having one full copy of your data on-site at all times provides a quick way to restore your data. However, you should also store your data at a different site or professional storage company that is registered to provide various levels of vaulting services.

For some businesses, there are strict regulations and guidelines that pertain to disaster recovery storage. Certain financial organizations are required to have at least two separate off-site storage locations. One of the storage locations is required to be a certain distance away from the primary business site. This distance is typically at least 60 miles. These strict requirements are in place so that a business can survive in the case of a nuclear, bomb, or natural disaster that could possible cover a large geographical distance.

Choosing an secure off-site storage location is an important discussion. It is best to go with a reputable, registered company that is certified to handle major disasters. For example, there are backup storage companies in the state of Florida that are registered to handle or survive various levels of hurricanes. In this case, choosing the storage vendor whose building could possible withstand the highest level of hurricane would be a good option.

ALTERNATIVE SITES

We have covered the importance of data backups and need to store your critical business data off-site. However, just as important, is the need to facilitate networks, systems, and other critical components that might be necessary in order to continue business related operations off-site in the event of a disaster. Simply put, if part or all of your building is destroyed as a result of a disaster, you will need a separate facility with the equipment necessary to get operational as soon as possible.

There are several options available for alternative or remote site operations. Again, budget, activation time response, and business continuity needs will usually determine the option utilized. There are also options available for making certain that you have the proper systems and components available in the case of emergency or disaster. Hardware manufactures as well as many local stores offer the ability to reserve specific inventory under a contract. This will allow you to have equipment such as servers and workstations available if needed. One thing to be wary of is that many companies rent the same space and/or systems to multiple businesses at the same time in the event of an emergency. Make sure that you deal with a reputable company or you may find yourself sharing the same building location and systems with multiple companies if a major disaster occurs!

The following are the most common implementations of alternative sites used for disaster recovery purposes.

Hot Site

This type of backup site is a site that can provide full business functionality in a very short time. It is a standby facility that has all the necessary equipment, software, applications, electricity, plus Heating Ventilation and Air Conditioning (HVAC) in place and ready to go. A hot site will usually only require the need for most recent data to be loaded to servers and workstations for full functionality. Mirrored transaction processing is typically used over high-speed transmission lines to keep hot sites up-to-date with current data and information. This type of site requires the identical security measures needed at the primary site. It is very expensive and requires constant administrative overhead to be functional.

Warm Site

This site is only partially configured. It will usually have all electricity and Heating Ventilation and Air Conditioning (HVAC) in place and ready to go. However, it might only house certain servers and specialized equipment. It is considered the medium level shell needed to get operations up and running. More servers, workstations, and equipment would be needed to get a warm site up to par for normal business operations. A warm site will usually have external network connectivity available. Again, the time and cost needed for recovery are the major factors here.

Cold Site

This site is a basic shell needed to get operations up and running. It will usually have all electricity plus Heating Ventilation and Air Conditioning (HVAC) in place and ready to go. A cold site doesn't have any equipment such as servers or workstations on site and there is usually no active connectivity to external networks in place.

DISASTER RECOVERY PLAN

A good documented disaster recovery plan is crucial to the survival of any business that wishes to provide quality service and remain functional before, during, and after a disaster has truly occurred. The goal of a solid disaster recovery plan is to provide proper policies, procedures, and documentation for backup and restoration of facilities and data in the event of an emergency. This plan should also include all emergency response procedures for all active employees who are considered part of a disaster recovery team.

A solid disaster recovery plan should be developed with the following considerations in mind:

- Create it with the intent to receive senior management support and "buy in."
- Minimize confusion in the event of an emergency.
- Protect company assets, information, and services.
- Minimize loss.
- Educate user environment through documentation and or training.
- Guarantee or prove the effectiveness of the DRP (Disaster Recovery Plan) through valid testing and simulation, if possible.

Once you have an approved DRP that is suitable for your company, it is important that it is kept up-to-date on a scheduled basis. A DRP from five years ago is not suitable or acceptable.

 For the exam, you need to know the first step when creating a disaster recovery plan. The first step needed when creating a comprehensive DRP is defining the goals that the plan will achieve. This will usually include the identification of what is considered a disaster or threat to your business.

An important fact to remember is that a disaster recovery plan is mainly concerned with procedures that reflect a company's ability to recover once a disaster has occurred. This is often referred to as *post-failure procedure*. *Business Continuity*, which will be detailed next, is concerned mainly with the guarantee that business operations will survive before, during, and after a disaster has occurred.

BUSINESS CONTINUITY

Business Continuity or *BCP (Business Continuity Planning)* focuses on preventing and protecting important business and financial assets in the case of a disaster. Through the implementation of approved processes and procedures, BCP attempts to prevent loss of mission critical services and provide functional fail over mechanisms if a disaster were to occur.

Technologies such as disk mirroring over the Internet, electronic vaulting, and Hierarchical Storage Management (HSM) can be implemented, at varying speeds and costs, to ensure Business Continuity during a disaster.

BCPs typically include a subplan, which is called a *contingency plan*. Simply put, a contingency plan is concerned with external actions and events that pose a danger to normal operations.

Once again, in order to define and implement a solid BCP, the first step is to define and document the goals that the BCP are expected to achieve. This includes identifying which of the company's functions are essential to daily operations and at possible risk and also getting management to budget according to these needs.

Other important considerations to keep in mind when developing a BCP/Contingency plan might include the following:

- Responsibilities checklist that includes all responsible members of the BCP team. This should include contact phone numbers, and define who will do what and where they will do it.

- Notification or alerting customers and normal employees that an emergency or disaster has occurred and that the plan is commencing.
- Damage assessment, control, and containment procedures.
- Recovery of critical systems.
- The ability to salvage a primary site or gain access to a remote backup facility or site.

FAULT TOLERANCE AND HIGH AVAILABILITY

In order for a business to provide *high availability* of an application's services and systems to its customers whether it is during or after a disaster has occurred, or for just plain daily operations, it is imperative that a functional network includes fault tolerant systems. *Fault tolerance* is defined as the ability of a program or system to remain functional in the event of a hardware or software malfunction or failure. There are various levels of fault tolerance that offer different levels of protection for systems and data. Whichever fault tolerant system or technology is at your company directly impacts the effectiveness of your disaster recovery plan. In other words, your DRP should always include tested fault tolerant systems.

RAID

RAID (Redundant Array of Intelligent Disks) is one of the most popular means of providing fault tolerant systems in use today. Through a process known as disk or data striping, RAID divides data into separate units and distributes the data across two or more hard disks. There are many variations of RAID available, the most popular are as follows:

- **RAID level 0:** This level of RAID is not considered fault tolerant. It spreads data in blocks across multiple disks but provides no data redundancy. This level of RAID produces better performance only. If one disk fails with this configuration, all data is lost.
- **RAID level 1:** This level is also known as *disk mirroring*. With RAID level 1, all data is duplicated or written to a second hard disk. If one of the disks fails, the information is still available on the second disk. This level of RAID is fault tolerant although its performance is not rated as well as RAID level 5.
- **RAID level 3:** This level also spreads data units across several disks but it also uses a dedicated disk for *parity information*, which is used for error correction purposes. In simple terms, it provides a basic level of fault tolerance.

- **RAID level 5:** This level provides excellent fault tolerance and good performance. It stores parity information across all disks in the disk array and provides concurrent disk reads and writes. It is the most popular RAID implementation.

Server Clustering

Another form of fault tolerance that provides high availability of services, applications, and resources is called *server clustering*. Server clustering is the grouping together of independent servers into one large logical system. Many modern day operating systems such as certain versions of Microsoft Windows and Linux offer the ability, through software, to implement server or resource clustering. Clustering is also known to provide *parallel processing* and *load balancing*. Parallel processing is the separating of process instructions so that separate processors can process them. This allows programs to run better and faster. Load balancing is the dividing or separating of work between multiple systems in order to process data and work loads more efficiently.

Another name sometimes used for a cluster of servers is a *server farm*. Again, typically, a server farm is a centralized group of systems or servers that act together as one unit to provide processing services such as authentication, backup, file, and print services, load balancing, and other resource sharing. A *Web farm* is a group of Web servers that provide Web pages and services.

Finally, servers systems are often made redundant as a method of fault tolerance. The concept of redundant servers is simply the application or mirroring concepts to two or more server systems. When data is written to one server, it is also *mirrored* or written to the second server. The implementation and utilization of a remote, redundant server is an excellent real time solution for any disaster recovery of business continuity plan.

POLICIES AND PROCEDURES

In most security conscious businesses, policies and procedures are implemented to provide a set of rules and standards for employees that represent management philosophies and opinions. Most company policies include certain sets of guidelines, standards, and procedures that should be implemented, enforced, and updated continuously to reflect changes in management wishes and direction properly.

SECURITY POLICY

A *security policy* is a detailed document that simply specifies how an organization will protect the business resources and assets.

 A security policy is a type of written policy that is never completed. It is a living document that deserves and requires continuous updates that reflect changes occurring over the lifetime of a business.

The security policy creation process typically begins by assessing the threats or risks that exist and developing a response team that will be implemented and empowered to respond to security related threats and issues. Next, usage policy statements are often created to identify the roles and responsibilities of specific employees. All company employees should be educated on the importance of the security policy and their roles regarding the security policy. Other statements are often added to security policy such as vendor or partner usage statements that specify the roles and responsibilities of clients and other third parties.

A good security policy might also include prevention statements and restorations statements that identify specific means that will be used to minimize security risk and steps that will take in response to security breaches. A very informative description of network-related security policy practices is available to the public at *http://www.cisco.com/warp/public/126/secpol.html#1a.*

Most security policies include an "Acceptable Use" clause or policy, which will be described next.

ACCEPTABLE USE

An Acceptable Use Policy (AUP) is a written and signed agreement that is usually required by a human resource department when a new employee is hired to work for a company. It can contain statements that reflect the company's policy regarding proper use of building facilities, equipment, software, applications, and other assets. If an employee breaks the rules included in the companies AUP, the employee can be subject to disciplinary action such as termination.

Many companies such as software manufacturers, Internet service providers (ISPs), network and online institutions require the signing of an AUP before access to their products or services are granted. Signing or agreeing to the terms included in an AUP means that you accept the rules and regulations associated with the AUP. Noncompliance with the AUP typically results in withdrawal or denial of the product or service and can result in legal action if deemed necessary by the owner.

DUE CARE

Company directors, managers, network administrators, and security personnel as well as other personnel are expected and required to carry out their specific job functions and exhibit due care. Due Care is considered acting in good faith as any "normal," "prudent" person would in a similar position or function. It is carrying out one's responsibilities with the overall welfare of the business or enterprise in mind.

Certain positions within companies require that key personal receive specialized training with certain business concepts and etiquette. Recent history tells us that high-ranking business representatives that do not act with due care can be legally held responsible for their actions.

Due Care and Due Diligence often are used synonymously; do not let this confuse you if you happen to see a question on the Security+ exam that references Due Diligence.

Due Care is concerned with maintaining confidentiality, integrity, and availability, which will be explained in the next section.

PRINCIPLES AND MANAGEMENT CONCEPTS

The development, implementation, and education of employees regarding company policies, procedures, and practices by business managers and leaders are imperative to the success of any productive secure business environment. There are also three very important management concepts that you should be aware of. They are as follows:

- **Confidentiality:** Ensures that only authorized individuals have access to services recourses and important data. It is critical that controls such as authentication mechanisms, firewalls, and cryptography practices are put in place to ensure confidentiality.
- **Integrity:** Ensures that important data remains unchanged or is not modified from any other state than is expected.
- **Availability:** Ensures that data, resources, and services are available to those who need them. In order for a business to be productive and profitable, it needs to ensure its employees, partners, and customers to have access to certain information and products. For

example, if a business provides a product for sale on an Internet site, and that particular Internet site is down, it is unavailable. DoS (Denial of Service) attacks on servers that provide Internet sites and service are often responsible for lack of availability.

 The security management concepts confidentiality, integrity, and availability make up what is well known in security circles as the *CIA Triad*.

SEPARATIONS OF DUTIES

Concerning information and communication security, *separation of duties* is the segmenting or the division of job responsibilities regarding highly sensitive information technology and data so that no one can solely jeopardize or compromise the integrity of data, a system, or a network.

Every job function in a company should have a documented job description that identifies the particular jobs requirements, responsibilities and duties. These job descriptions should be updated whenever an employee leaves the company, is terminated, or moves to another position within the company. This documentation should also be updated as job requirements change.

SLA (SERVICE LEVEL AGREEMENTS)

SLAs are agreements or contracts between vendors of services or products that specify what the service agreement will provide. A good example of an SLA would be a contract from an Internet service provider who, in writing, guarantees a certain level of access speed to the Internet.

Most SLAs provide statistics that can be used to compare services with other vendors of a similar product. For example, companies that provide access to the Internet often include such metrics as measured response time, help desk response time, total number of users who can be signed on at a time, accessibility time, and other information that can be used as a benchmark.

Today, many enterprise-class companies require SLAs and other contractual agreements to be in place with their service providers in order to guarantee a certain level of access and integrity to business customers.

DISPOSAL/DESTRUCTION

The proper handling of confidential company information, whether it be hard or soft copy, should always be a consideration when developing a company security policy or just handling sensitive information. Rules should be put in place and these rules should be brought to the attention of every employee that handles important and confidential data. Most companies have a disposal policy that states all paper and printed output must be shredded in a crosscut shredder before it is recycled or thrown away. Most of these policies are not limited to paper output. They also include policy that pertains to the erasing of data stored on media such as floppy disks, CDs, zip disks, hard drives, and other forms of media storage before the media is thrown away, recycled, donated, or sold to a third party.

Information theft is not isolated to online sources. A popular technique known as *dumpster diving* is commonly used by hackers to obtain information that can be used to gain access to systems and confidential company information. In simple terms, dumpster diving is going through someone else's trash with the hopes of finding information such as names, ID's, phone lists, passwords, network information, PINs, account numbers, and other information that can be used for social engineering attacks and access to information systems.

HR POLICY

Human Resource departments play a very important role in the administrative management of security practices and policies. These departments should have strict HR (Human Resource) policies in place in order to ensure the proper procedures are implemented and followed regarding the hiring and firing of potential and company employees. HR personnel are also required to use due care and follow a proper code of ethics when managing the human factor of an organization.

HIRING

Hiring quality, responsible, trustworthy employees who meet a business's needs is an essential process in protecting a company's infrastructure. Proper HR policies and procedures should be in place when recruiting new employees. HR personnel should be trained and familiar with hiring laws, regulation, and company policy.

Background checks should be conducted to ensure that perspective employees meet policy and company needs. Multiple references should be consulted to ensure a possible employee's integrity and honesty.

Once hired, it should be policy that new employee's job responsibilities are clearly defined and presented to the employee. The employee should be well educated on the rules, regulations, and policies associated with the company and the employee's role in the company. Building and network access should be granted to new employees based on their roles, applying the principle of *least privilege*. In other words, give an employee access to physical places and logical resources needed to carry out only their specific job functions.

TERMINATION

As is the case with hiring practices, employee termination policies require great attention to be effective. All of the same practices mentioned earlier should be considered. HR personal must be educated with laws and regulations as they apply to the dismissal of company employees. Documented company policy as it applies to termination should be strictly followed.

When an employee is terminated or leaves the company for any reason, all responsible management, security, and network personal should be notified that the employee has left. All physical access privileges to the building and other secured areas should be immediately revoked for the employee. All network and resource access should be immediately disabled or removed for the former employee. If the former employee requires access to personal property before leaving the business, they should be escorted by HR and security personal to and from all required locations.

CODE OF ETHICS

A *code of ethics* is a documented representation of the values associated with a profession or business. They are the collaborative beliefs or values associated with job responsibilities that are used to assure employees, partners, and clients that a company, its management, and its employees will act responsibly and be accountable for their actions as they pertain to the products they provide and the customers they support.

A solid company code of ethics typically will include the following abilities:

- Employees can monitor the behavior of coworkers as the behavior applies to the company's documented code of ethics.
- Management and HR can educate and carry out the values contained in the code of ethics.
- Management and HR can dismiss or reprimand employees for not abiding by the company's documented code of ethics.

Just about every job or industry that provides a product or service has as associated code of ethics. Even the Internet has a code of ethics; use of government information and Web sites have an associated code of ethics; and yes, even certain security certification requires a code of ethics. Currently, those who are CISSP (Certified Information Security Professionals) certified are required to follow the code of ethics stated at: *http://www.isc2.org/cgi-bin/content.cgi?category=12.*

INCIDENT RESPONSE POLICY

An *incident response policy* is a set of instructions, guidelines, and rules usually created by a CIRT (Computer Incident Response Team), which are most often a combination of management and a skilled team of those who are technically inclined. The incident response policy should be followed if a known emergency, security-related incident or disaster has occurred. Most well-organized and well-documented incident response policies will state or define what an *incident* really is to the company. This is usually the first step in the development of an incident response policy. The policy will list the emergency contacts and prioritize the order in which these contacts are notified. The policy will also require that an incident response report be filled out if a security issue or emergency has occurred. An incident response report will typically include items such as the following:

- Time and place the incident occurred.
- Summary of people and systems that were involved with the incident.
- Description in detail of the incident.
- Documenter's name, the time the report was initiated, and the people who were notified when the issue occurred.

Proper incident handling includes three important factors. They are incident reporting, incident analysis, and incident response. For security issues and emergencies to be properly taken care of, and the earlier mentioned important factors to be addressed in a controlled manner, it is critical that incidents be reported as soon as possible. Most computer security incidents require quick action. If an intruder, virus, or disgruntled employee is not handled with swift action, devastating results to company equipment and data may be the end result.

The Carnegie Mellon University provides the following very informative Web site that details incident response handling and services: *http://www.cert.org/csirts/csirt_faq.html.*

PRIVILEGE MANAGEMENT

Privileges are defined as user rights that are generally granted to specific users or groups of users within a specific system or network. These rights allow users or groups of users to carry out specific operating system tasks such as the backing up a system, shutting down a system, or changing a systems time. Figure 6.1 displays the many local user rights assigned within the Windows 2000 Professional operating system.

USER/GROUP/ROLE MANAGEMENT

The concepts of managing users and groups can be best explained by describing the basic administration of users and groups inherent with the Microsoft operating system, Windows NT.

A Windows NT workstation can participate in a workgroup or domain environment. In a workgroup, each individual computer system houses its own SAM (Security Accounts Manager) database. In a domain model environment, the SAM database is located in a more central location such as a PDC and BDC. This allows administrators to control user access to the network as well as provide the sharing of network resources from a centralized location.

A *work group model* mirrors a peer-to-peer network in which security and the sharing of resources is controlled at every individual machine. Imagine organizing a work group of 200 users. You would have to control user access to the workgroup and password protected shares at every single system!

Windows NT comes with an administration tool known as User Manager. User Manager has several built-in groups. Figure 6.2 displays User Manager. These groups are designed for ease of administration. A Windows NT built in-group has preassigned user rights. Windows NT user rights allow users or groups of users to carry out specific tasks such and the right to backup the system, shut down the system, or change the system time. After a user ID has been created in User Manager, it can be placed in a group. When the User ID is placed in the group, the ID inherits all of the rights associated with that group. For example, if the user ID JSHMOE were created and added to the Administrators group, the user JSHMOE would inherit all of the user rights associated with the Administrator group.

Every Windows NT workstation or server has a set of built-in local groups. If a user has been placed into a local group, it is possible for the user to access resources and be granted rights on the local system. To ease

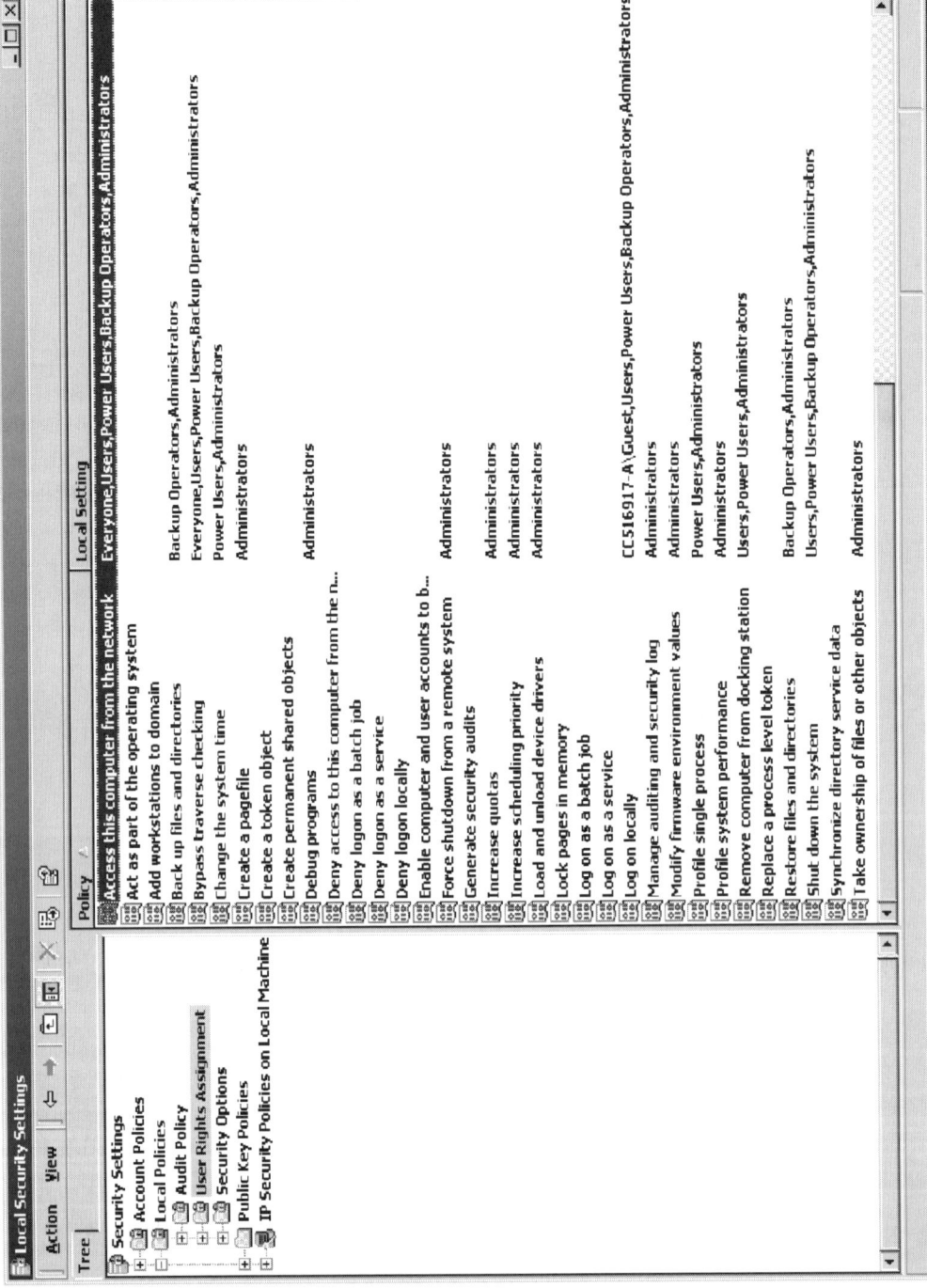

FIGURE 6.1 Windows 2000 Professional local rights.

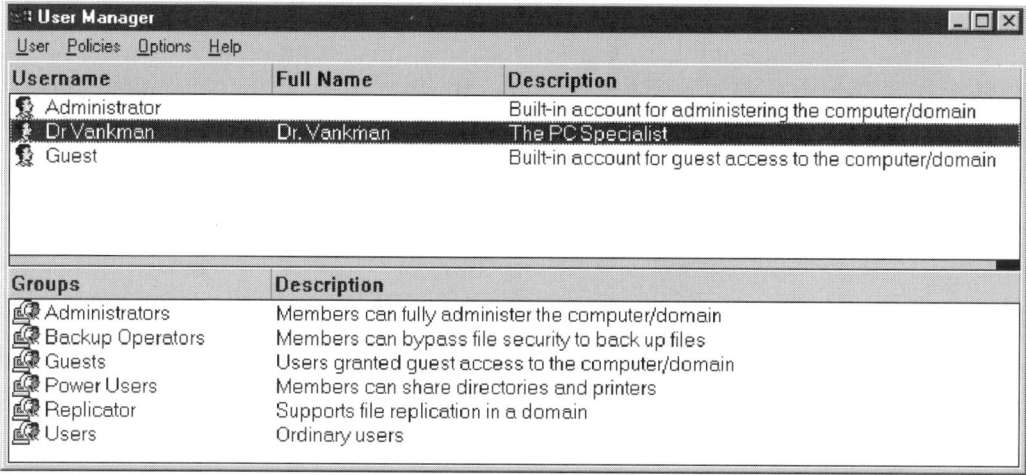

FIGURE 6.2   The Windows NT User Manager.

domain-level administration efforts, Windows NT Server also makes use of global groups. Simply put, many users can be placed into a Global group. The Global group can then be added to a local group located on a workstation or server. The end result is that all users in the Global group can access resources assigned to the local group on a particular workstation or server.

In addition to creating user IDs and the assignment of user rights, User Manager also has the ability to audit the success and failure of events that occur on the system. An administrator can audit access to files and objects, users who have logged on or logged off of the system, and security policy changes, just to name a few. The results of the events that have been audited are displayed in the Windows NT Event Viewer.

The exam is likely to challenge you with the various types of access control models, permissions, and roles implemented with the access control types mentioned in Chapter 2. These include Role-Based Access Control (RBAC), Lattice-Based Access Control (LBAC), Task-Based Access Control (TBAC), and Terminal Access Controller Access Control System (TACACS). This is a good place to go back and refresh yourself. If you do not know these well, you will probably not pass this exam.

NOTE

LSA, SID, AND ACL

Security access to all resources and the entire security sign-on process that takes place behind the scenes when a user logs on to a network running an operating system server such as Windows NT is beyond the scope of this book. However, a basic explanation and several important terms are in order to provide you with a general understanding of how this process works. This might assist you with the basics of understanding general authentication practices.

When a user logs onto NT workstation, the LSA (Local Security Authority) generates what is called a *security access token* or *SAT* for short. This SAT is assigned an SID (Security ID) for the user. The unique user SID contains access rights and privileges that have been assigned to the user's ID that was created in User Manager or User Manager for domains. Windows NT maintains what is called an *Access Control List* for all objects that exist in the Windows NT domain. An object can be a file, folder, or printer share, just to name a few. In order for a user to be granted access to an object on the domain, the user's SAT must be accepted by the ACL.

SINGLE SIGN-ON (SSO)

Single Sign-On (SSO) user authentication is the use of one username and one password in order to give the user access to all resources, applications, and other shares that the user has been given permission to use. In other words, a user only has to sign on one time to an authenticating server. From that point on, permissions are used to grant access to the user for various resources so that the user does not have to keep entering passwords.

The Single Sign-On model is popular in most client/server network environments. Server operating system software such as Windows NT, Windows 2000, and Novell are based on a Single Sign-On authentication model. This allows ease of user administration by allowing users to be placed into groups. The groups are then assigned permissions to recourses as opposed to assigning access permissions to each individual user.

The main benefits of implementing Single Sign-Ons are as follows:

- Users only have to sign on one time with a primary server or in domain controller in order to access enterprise-wide recourses if needed.
- Risks involved with individual users having to remember multiple user IDs and passwords are eliminated or reduced.
- Administrative overhead is reduced. Administrators don't have to waste precious time resetting passwords on shares because users lost or forgot them.

The inherent threat that is associated with the use of SSO is that if a hacker or attacker gains a single user ID and associated password, an entire network can be compromised.

Two-Factor SSO is often a better way to go in order to provide better security in single sign-on environment. With Two-factor SSO, a user provides an ID and a password combination and is also required to authenticated with a token or biometric device. Most password cracking programs are written with the targeting of SSO operating systems in mind. Use Two-Factor SSO for better security if possible.

Centralized/Decentralized

In the world of security policy and practices there are two main types of access control environments: *centralized* and *decentralized*. Centralized access control security practices focus on maintaining and storing user IDs, passwords, permissions, and access rights in a single centralized location. The decentralized access approach to network security focuses on the control and storage of user IDs, passwords, permissions, and access rights across an enterprise in multiple locations. Most modern day client/server networks implement the decentralized approach.

Security Roles and Responsibilities

The Security+ examination might question you on the following security roles and responsibilities. These are very general descriptions that are common to most businesses as well as most information security related exams:

- **Management:** Responsible for ensuring that all employees follow security policies and practices. Responsible for protecting all company assets. Responsible for exercising due care in all of their affairs.
- **Owner:** Responsible for the protection of company information. The owner's responsibilities include determining the classification level of data, making changes to the classification level, and assigning or delegating who will be responsible for data and the security of data. (Data classification levels are described under the "Documentation" section later in this chapter.)
- **Custodian:** This is a delegated position that is responsible for the protection and testing of data and procedures. Custodian duties include the verification of system backups and restores as well as keeping updated documentation that supports testing and production efforts.
- **User:** Anyone who uses data routinely to carry out their normal job responsibilities. Users should be responsible and accountable for the data they work with.

FORENSICS

Computer *forensics* can be described easily as an investigation of a computer network or system in hopes of recovering data and other information that can be used as evidence to properly prosecute the sources responsible for a computer-related theft of crime. Computer forensics is concerned with how laws apply to computer science and technology in general. Its focus is on the collection and preservation of computer-related information and evidence in order to prove that a computer crime has occurred.

In today's electronic world, most information is stored on media such as hard drives, CD-ROMs, floppy disks, and other forms of electronic storage. The extraction processes that take place to find hidden file formats, secrets, files that have been encrypted, and other evidence from these types of media are the main focus of computer forensics.

Typically, a third party is used to provide the recovery and extraction of information from a system that has been damaged. Special software and the ability to find traces of malicious activity very quickly are the skills and tools held by most third-party computer forensic specialists. Most network administrators carry out computer forensics techniques on a daily basis but don't necessarily call it forensics. They call it *auditing, logging, data backup*, and *data restoration*. You will have to know what forensics is for the exam. However, it is likely that you won't get hammered with questions concerning computer forensics. This is a fairly new area that will be targeted more heavily in future security related examinations.

EVIDENCE COLLECTION AND PRESERVATION

The proper collection, storage, preservation, and protection of evidence that has been identified on media or a system that has compromised are critical to the evidence life cycle concerning computer forensics.

If potential evidence is corrupted, damaged, or not handled with Due Care, the evidence might not be admissible in a court of law. The following items should be considered when collecting, handling, storing, and transporting possible computer-related forensic evidence:

- **It is imperative during the collection stage of the forensic process that no information is damaged, misplaced, or destroyed.** In other words, if you are investigating a compromised system and manage to delete or modify code or information that is considered evidence, you might have just ruined your chances of providing accurate evidence. Another consideration is that you must not further infect a possibly compromised system. When considered compromised, a system should be disconnected immediately from a network and other resources. If a system has been purposely infected with code or a virus from one source and is further infected during the investigation process, it is possible that any evidence obtained will not be admissible for legal proceedings.

- **The system or media being examined must include documented information pertaining to file structure and all other identified system information.** The documentation must contain the signature of the person examining the information and the date it was examined.

- **Once the evidence has been extracted and gathered, it is of utmost importance that it is properly handled, stored, and packaged.** Rubber, static-free gloves should be worn when handling evidence so as to avoid further fingerprints. The evidence should be stored in a sealed container. The container and all pertinent information must be labeled and documented. This documentation must include the signature of the person carrying out the investigation and the date the investigation took place.

- **Evidence should be transported to a third-party storage facility until it is presented as legal evidence in a trail or proceeding, or is returned to its original location or owner.** It is important that the same level of care is taken when transporting the evidence to a legal proceeding as it was taken when the evidence was stored originally.

- **A chain of custody must be in place to ensure that it is always known where the evidence is physically located and who has possession of it.** A secured logbook should be used as a verification and identification tool during this process.

The National Institute of Standards and Technology is an excellent site that has a wealth of information regarding Forensics. For more information, visit the NIST at *http://www.nist.gov/*. For another great explanation of Forensics visit *http://www.forensics-intl.com/def4.html*.

RISK/THREAT IDENTIFICATION

Risk management is the process or processes that a company or enterprise implements in order to reduce loss of assets or financial standings that can result from theft, accidents, or lack of proper management Due Care and responsibility. In simple terms, risk management focuses on the reduction of threats to a company's assets. For more specific network and security related study needs, it can be said that risk management focuses on the implementation of information security practices in order to identify possible hardware/software threats and vulnerabilities that exist both inside and outside of an organization.

In order to prevent or offset risk and the possible financial losses that exist if proper prevention methods are not put in place, one must identify what real risks and threats exist, know what assets are at risk, and analyze what, if any, controls should be put into place to avoid loss. Identifying risks includes the following:

- Knowing the actual threats that exist.
- Knowing the possible consequences or repercussions that exist.
- Knowing the possible likelihood that an event or disaster will take place.
- Knowing if the event or disaster will reoccur and at what frequency.

Risks can be categorized or isolated to give you a better understanding of how to identify them. The following are categories of risks or threats that could possibly exist:

- **Information warfare threats:** Terrorism using Information Technology.
- **Data threats:** Viruses or malicious code.
- **Personnel or employee threats:** Unauthorized physical or virtual access.
- **Criminal threats:** Theft or vandalism.
- **Environmental threats:** Natural disasters and facility-related failure.
- **System/Computer:** Hardware/software failures.

After you have identified all risks that present a threat to your security, it is important that you evaluate what systems, people, and other assets are vulnerable to these risks. Once vulnerabilities are identified, a control process can be implemented. Once again, the type of controls, insurance, and safeguards are all determined by the mighty dollar.

Management buy-in concerning the protection of an organization against threat is a must.

RISK ANALYSIS/ASSESSMENT

Risk analysis includes identifying important assets and identifying possible risks to these assets. Risk analysis also includes implementing safeguards to prevent or offset the risks or threats that exist.

There are three major items you should be familiar with when preparing a risk analysis. You should be able to estimate the possible losses that could occur, analyze/assess potential risks, and be able to produce an Annualized Loss Expectancy report (ALE). In order to produce an annualized loss expectancy figure, and for the exam, you should know the following:

- **SLE (Single Loss Expectancy):** This is the expected financial loss due to single failure or event.
- **ARO (Annualized Rate of Occurrence):** This is the estimated amount of times or frequency that the failure or event will occur within a year.

In order to produce an ALE, you must multiply the SLE by the ARO.

 ALE = SLE × ARO. You should know this formula for the exam.

The ALE should include a list of all assets, all possible threats, the potential for threats, the financial and physical loss that can occur from these threats, and recommended remedies to reduce the risk potential.

There are several types or approaches of risk analysis that you should familiar with for the exam. They are as follows:

- **Quantitative risk analysis:** This type of risk analysis is based on two factors—the probability that an event will occur and an estimate of the amount of loss that will result. This type of analysis is less popular than qualitative risk analysis because it is based on data that is not factual. In other words, the probability that an event will occur cannot always be accurately measured. With this type of analysis, the risks are guessed and money is appropriated as a means to offset or take care of the aftermath if an event should occur.
- **Qualitative risk analysis:** This type of risk analysis is much more popular. It is based only on potential loss not probability. With a qualitative risk analysis, threats and vulnerabilities are analyzed and defined. Then, controls are put into place to offset the risk. These controls are deterrent, preventive, corrective, and detective.

EDUCATION

Educating all employees, contractors, and other third parties about an organization's security practices, policies, and procedures should be a priority. It is the responsibility of management and human resources to see that all communication, training, documentation, and practices are brought to the attention of employees on a need-to-know basis. All the physical and logical security implementations as well as documented policies are going to do you little good if your general work population is not educated on the fundamental rules and regulations that exist within your organization.

Most business today invest in specialized third-party security training for authorized personal. They also require Human Resource departments to provide mandatory educational meetings on company policy, handbooks, general security, and safety topics. Educating employees and partners with up-to-date security practices can be very beneficial to the overall health of you organization.

Security education and training should be paid for and provided at all levels of an organization. Managers should receive security management training; technicians should be required to take CISSP and Security+ training courses as well as become certified.

If your company is unable to send you for training, CBT (Computer Based Training) courses can be made available or interactive security training over the Web can be sought after.

Some of the benefits that can be seen as a result of security education are an overall increase in productiveness, a decrease in theft, fraud, vandalism, and unauthorized access. Recent world events tell us that emergency planning, disaster recovery, and security training and education should be among our top priorities.

COMMUNICATION

Communication is the key to educating users and employees with updated security policies and general security awareness. If you don't educate and communicate new policies and procedures to fellow employees, they obviously might not know what to do in the event of a disaster or security related incident. Regularly scheduled meetings should take place to keep users informed of updates or newly introduced policy and general company handbook updated. Documents should also be distributed outlining any policy changes or updates. Communications tools such as reminders, posters and intranet Web site scan also be used to update or educated employees on current practices.

DOCUMENTATION

The documenting of important company data, procedures, policies, and other vital information is critical to the life cycle of any healthy business. It is important that company data and information be prepared, classified, handled, stored, retained, and disposed of if necessary with great care and instruction. It is likely that you will see many questions on the exam that focus on the proper methods of handling, retaining, and destroying data and important information. Place your study focus on those subtopics in your final exam study preparation of documentation.

Standards, Policies, and Guidelines

Standards are approved, accepted, and defined mandates or rules that apply to activities and functions. An example of a standard is that all people are required to sign in with a security guard, or all employees must be drug tested before being hired. Standards are implemented to support organization or company policy. Many technical and security related organizations provide standards for compliance and acceptance. In other words, in order to receive their services, you must follow their rules or standards.

Policies, which are supported by standards, are typically the collaborative ideas and viewpoints of management, owners, and partners. These policies are intended by management to be used as a road map for employees to follow. In other words, management creates the policy and employees are expected to adhere to these policies, or management viewpoints. In order for the policy to be effective, it must be followed and enforced with punishment if necessary. Policy must also be kept up-to-date and communicated to all employees who it affects.

Guidelines are general courses of action that can be taken or followed based on standards and policy. For example, a company might have a policy that states, "Smoking in designated areas only." The guideline for this policy might include the actual designated areas and other information.

Systems Architecture

Documenting a system or network's components and software architecture is a critical part of overall security as well as for the implementation of new code or other components. The need for better documented system architecture and a better communication level between manufactures, programmers, developers, network engineers, and security analysts has become very apparent as technology has progressed. If a system's architecture is not well known and properly documented, the result of a project will typically be failure.

The proper documentation of system architecture should include a "how-what-why" approach—How the documentation serves it intended audience; what is the actual architecture of this system; and why the architecture has been designed the way it exists.

Documenting a system's architecture provides a set of guidelines or views that can be followed and addressed when new implementations are considered. These views provide a framework that can be used to manage the effectiveness of a project. They can also serve as signal light that determines whether the project should continue or not.

Change Documentation

Change Documentation is needed to preserve integrity to a program, network, system, or business when changes are needed and made to the configurations, policies, or documentation in general. For example, a change control document might exist to provide a structured approach as to how procedures exist for moving test data and programs into a production or a live environment.

Change control procedures are a means of tracking, auditing, and controlling all changes that are made to a particular system or program. The following five accepted procedures are considered the general guidelines that should be followed when implementing a change control process:

1. Applying for the introduction of a change.
2. Cataloging the intended change.
3. Scheduling the change.
4. Implementing the change.
5. Reporting the change to all parties involved.

These additional items should be considered concerning the implementation of a change to a production environment:

- Before the change is moved to production, it is imperative that it has been tested in an orderly and acceptable fashion.
- Users and affected parties must be informed of the change.
- The effects of the change must be analyzed after the change has occurred.

In very simple terms, change control process and sign-off are needed to reduce the risks associated with the implementation of new applications and products into a production environment. In real life, new code and application integration typically goes through the four following processes or phases:

- Development
- Testing
- Implementation
- Review

Data Classification

Classifying data into separate distinct categories is used as a method to comply with company, state, local, and federal regulations. Financial and other accounting businesses as well as many other companies that provide products and services are most often required by law to classify data.

A data classification plan or scheme is important because it helps identify companies' important assets; identifies how data is protected; and provides a means of demonstrating an organization's commitment to security.

Data should be classified by certain criteria. The following criteria are accepted practices for classifying data:

- **Data value:** How valuable is the data to the company, partner, or client?
- **Data age:** How old is the data? The data value might decrease as the data gets older.
- **Useful life of data:** The useful life of the data might change over time. If the data is no longer needed, it might be reclassified or declassified.
- **Personal or job association:** Data might be identified with a specific person or job function.

There are two main categories of data classification: *commercial* and *government*.

 Know the data classifications for the Security+ exam. It is very likely that you will see questions that relate to them.

Commercial Data Classification

This category consists of the following:

- **Public:** This data should not be readily available to the public. However, if the general public views this data, it will not cause damage.
- **Sensitive:** This data needs a high level of protection and should remain confidential at all times.
- **Private:** This data is personal and intended for business or company use only. Disclosure will generally not result in loss or damage.
- **Confidential:** This is the highest commercial classification. This is data that is highly sensitive and meant for internal business or company use. The disclosure of this type of data will result in extreme damage or loss.

Government Data Classification
This category consists of the following:

- **Unclassified:** This data is not sensitive in nature. It has no classification. This information is viewable by the public.
- **Sensitive but Unclassified:** This data is considered secret. However, if it is released to or viewed by the public it will not cause damage, harm, or loss.
- **Confidential:** Information that is considered Confidential should not be made public. Moderate to serious damage could result.
- **Secret:** Unauthorized disclosure of Secret information or data is likely to cause serious damage.
- **Top Secret:** This is the most classified of the classifications. Grave or extreme damage and/or circumstances can result from the disclosure of this information.

NOTIFICATION

As documented policies, standards, security procedures, technical information, and other important company information changes, it is important that the proper employees are notified. As mentioned earlier in the chapter, lack of communication can result in disastrous consequences. Some of the negative results that can occur by lack of due care concerning notification and communication are as follows:

- System failure
- Application failure
- Fraud

- Theft
- Disgruntled employees
- Lack of productivity
- Legal issues

Anything negative (within reason) that could possibly happen in an environment should be considered when documenting and providing notification. If company policy changes, the proper employees should be notified. If a disgruntled network administrator or security analyst leaves or threatens the integrity of the company, immediate notification to the proper personnel should occur. If system software discovers malicious code, an immediate alert in the form of an e-mail or a page should be sent directly to the system or network administrator. If the building maintenance crew needs to mop or clean a stairwell, signs or notification should be posted to avoid accidental injury.

Proper notification documentation should include emergency contact information for all company managers, security personnel, Human Resource personnel, network and disaster recovery teams (both local site and enterprise-wide if necessary), and building maintenance. It should include phone numbers for local fire, police, and hospitals. This document should also include whom to contact for what. There should be no question in the person's mind reading the document regarding whom to call. Emergency notification requires quick thinking and response. It is likely that this document will include personal information that should only be used by authorized personnel. It should be available on a need-to-know basis only.

INVENTORIES AND LOGS

All company assets should be inventoried in a control document. This inventory list should include such assets as licensed hardware, software, applications, products, office equipment, and all other company assets. Every department within an organization should be responsible for keeping their section of the company inventory list up-to-date and accurate. A specific person within each department should be assign with this responsibility. Serial numbers, proof or purchase, and asset numbers should be associated with the various company assets. The inventory list should be audited on a regular basis to account for all assets and verify the inventory list's accuracy and integrity. A duplicate inventory list should be stored off site in the event of an emergency that destroys the original list.

For ease of administration and the collaboration of company inventory, an inventory database software package should be purchased or created.

Just be sure that access to certain file or fields within the database are accessible only to those who need access.

System and security logs also require careful consideration and handling. Audit logs are often used to prove that illegal activity such as unauthorized access to confidential information has occurred. Security cameras and magnetic badge systems record and store pictures and logs that can be used as evidence in a court of law or be provided to insurance companies for reimbursement if something negative occurs.

The proper logging of certain system events and firewall activity are critical to the security welfare of a network. If logging is set up properly, hackers and malicious activity can be recorded and tracked. It is important to note that many hackers are well educated at erasing their tracks. They will often delete or remove log files that are typically stored in default locations after they are done doing whatever it is they do. Good security practices suggest that you should rename and move your log files on a regular basis to avoid this activity.

When creating a logging plan, the following items should be considered:

- What events should be logged.
- The storage location of the log files and an estimated file size of the logs.
- Who will require access to the log files.
- Whether the log file should be encrypted.
- Frequency at which the log files should be backed up and stored.

STORAGE/RETENTION

Log files and company inventory lists should be stored and retained according to company policy and possible regulations. It is important to note that certain operating system log files are by default limited to a certain sizes and retention periods. For example, by default, the operating system Windows NT limits the size of the system, security, and applications event logs to 512KB each and has a default retention period of seven days. With these settings, the oldest audited events will be removed from the system when the 512KB limit is met or if seven days pass. This is important to consider. You need to coordinate the backups of your audit trails and logs with the setting you have specified in the particular operating system you are using. Auditing and backing up everything all of the time costs money and hampers system recourses and response times. Make an organized plan. You might need to log files from six months ago in order to prove wrongdoing.

Important information (such as that contained in documents and backup media) that is considered valuable to a company should be stored at an off-site, professional, registered storage company. Access to the stored information should require proper identification and signatures where necessary. If your information is stored off site, you should make sure that you can readily gain access to it 24 hours a day, seven days a week.

DESTRUCTION

Every business should have a document prevention/destruction plan/ policy and associated guidelines to support the policy regarding the destruction of both physical and logical documents. Implementing such a plan and carrying out proper document disposal methods and techniques may save your company or organization from certain fraud, theft, espionage, and possible legality issues.

There are many certified third-party businesses that offer document destruction and the recycling of important and confidential information. Typically, these services provide on-site secure storage containers that confidential information can be placed in and sealed for further destruction and possible recycling. Do not ever place confidential company information in a physical public recycle bin. At the very least, your company should have a two-way shredder that enables documents to be destroyed on site.

As far as logical document destruction goes, do not ever assume that the top-secret document that you just placed in the Windows Recycle Bin has been deleted. If you empty the Recycle Bin, the document can still be recovered from the systems FAT (file allocation table.) If you need to keep and store the top-secret information, encrypt it.

In order to completely remove information from most electronic media, the media needs to be electronically or magnetically erased. For most, encryption is the way to go. For the military, they usually burn all confidential information that is no longer needed. For the exam, multiple reformats of the media are acceptable for most company reuse policies.

TEST TIPS

The following operational/organizational Test Tips are in place to give you a well-needed edge in the exam room. Focus on them well for they may be the keys to your success. Please remember that some of these tips are not mentioned in the chapter. Some of them come from hands-on experience; others are from important security conceptual references.

√ *Penetration testing* is used to see how vulnerable a current environment is to risk and vulnerability. This testing is often done with an attacker's perspective in mind.

√ In operational security management terms, the protection of Confidentiality, Integrity, and Availability make up what is known as the *CIA Triad.*

√ *CCTVs (closed-circuit televisions)* are often used in combination with surveillance cameras as physical security monitoring devices. CCTVs' signals use private channels to provide a signal from the TV to the camera. They do not broadcast signals to public areas.

√ The exam is likely to confront you with the proper suppression methods that should be implemented in case of a fire.

√ The best way to protect the information stored on portable systems such as laptop computers, PDAs and wireless cell phones is to use encryption if possible.

√ Secured doors should be resistant to forcible entry and should unlock automatically in the case of an emergency. It is very important that you know this for the exam.

√ The goal of a solid disaster recovery plan is to provide proper policies, procedures, and documentation for backup and restoration of facilities and data in the event of an emergency.

√ A *GFS* backup strategy provides the fastest and easiest restore.

√ In order to define and implement a solid *BCP*, the first step is to define and document the goals that the BCP is expected to achieve. This includes identifying which of the company's functions are essential to daily operations and possibly at risk.

√ Many modern day operating systems such as certain versions of Microsoft Windows and Linux offer the ability, through software, to implement server or resource clustering.

√ *SLAs* are agreements or contracts between vendors of services or products that specify what the service agreement will provide.

√ A *Disaster Recovery Plan (DRP)* focuses on the implementation of procedures that should be followed during and after a disaster. A *Business Continuity Plan (BCP)* focuses on prevention and how a disaster affects the overall business plan of an enterprise.

√ Building and network access should be granted to new employees based on their specific roles. The principle of *least privilege* should be considered with new as well as existing employees.

✓ When an employee is terminated or leaves a company, all physical and logical access should be denied for the individual.

✓ *Two-Factor SSO* is often a better way to go in order to provide better security in Single Sign-On environment. With Two-factor SSO, a user provides an ID and a password combination and is also required to authenticated with a token or biometric device.

✓ *Change Documentation* is needed to preserve integrity to a program, network, system, or business when changes are needed and made to the configurations, policies, or documentation in general.

✓ A *chain of custody* must be in place to ensure that it is always known where the evidence is physically located and who has possession of it.

✓ If potential evidence is corrupted, damaged, or not handled with *Due Care*, the evidence might not be admissible in a court of law.

✓ With *qualitative risk analysis*, threats and vulnerabilities are analyzed and defined. Then, controls are put into place to reduce or offset possible risks.

✓ With *quantitative risk analysis*, risks are guessed and money and/or insurance are appropriated as a means to offset the risk.

✓ Proper *notification documentation* should include emergency contact information for all company managers, security personnel, Human Resource personnel, network and disaster recovery teams (both local site and enterprise-wide if necessary), and building maintenance.

✓ You will be tested on the proper document handling and document destruction basics. Know them.

CHAPTER SUMMARY

Recent historical events have brought to the forefront the importance for such things as physical security, disaster recovery planning, risk identification, and education. The effects on particular business operations and an economy as a whole can be severely impacted if proper measures are not taken to minimize the physical and logical threats that are ever increasing in today's world.

Most of the concepts and technical information stressed in this chapter have been around for quite some time. It has become an industry demand for properly educated certified professionals to implement these policies and procedures to better ensure the operational continuation and health of businesses.

At this point, you should at the very least be able to handle questions that relate to the following topics of discussion covered in this chapter:

- Physical security implementations and countermeasures.
- Aspects of Disaster Recovery Planning, procedures, and backup types.
- Business Continuity Planning and implementations.
- Forensics and the importance of gathering evidence that can be used to prosecute if a computer-related crime has occurred.
- Specific management/company policies and guidelines that should be documented, implemented, and followed by company employees.
- Ability to identify certain risks that are considered a threat to normal operations.
- Importance of educating all company employees regarding proper policies and procedures.

REVIEW QUESTIONS

1. **John Doe from the local power and light company calls you and requests access to your secure remote location for a routine meter reading. What do you do?**

   ○ A. Let him in. All local regulations give power and light companies access precedence for routine maintenance.

   ○ B. Ask the guard at the front desk what he thinks you should do.

   ○ C. Check your documented access control policy, which might contain an access list. Let him in if he's on the list. Deny access and call the power company if he is not.

   ○ D. Tell him to say "Open Sesame."

   Correct answer = C

   The answer to this question is based on what your documented access control policy states. In other words, you might have a digital pass system, an ACL (Access Control List), a remote surveillance device, or biometric device that can be used to determine remotely whether the worker should be allowed or denied entry to the remote site. All other choices are invalid.

2. **Which of the following choices is not a considered a physical security control?**

   ○ A. Cryptography.

   ○ B. Mantrap.

   ○ C. Turnstile.

   ○ D. Biometric device.

   Correct answer = A

   Cryptography is used to transform or encrypt plain text into an unreadable or unidentifiable format known as ciphertext. In order for the encrypted text to be understandable, it must be decrypted. Cryptography is not considered a physical security control. Mantraps, turnstiles, and biometric devices are considered physical security controls.

3. **Which of the following choices represent preventive control measures?**

   ○ A. Implementing more restrictive file level permissions after a breach.

   ○ B. Antivirus protection and strong passwords.

   ○ C. Reviewing log files and monitoring suspicious activity.

   ○ D. A good backup system.

   Correct answer = B

   Implementing antivirus protection and strong passwords are examples of preventive controls. Implementing more restrictive file level permissions after a breach is considered a corrective control. Reviewing log files and monitoring suspicious activity are considered detective controls. A good backup system is a recovery control.

4. **A security management concept known as the CIA Triad represents what?**

   ○ A. Confidentiality, IDEA, Auditing.

   ○ B. Customers, Integrity, Auditing.

   ○ C. Conspiracy, IDEA, Availability.

   ○ D. Confidentiality, Integrity, Availability.

   Correct answer = D

In operational security management terms, the protection of confidentiality, integrity and availability make up what is known as the CIA Triad. All other choices are invalid.

5. **What is best way to secure portable devices that contain confidential information?**

   ○ A. Use a BIOS-protecting password.
   ○ B. Secure the devices by storing and locking them in cars, vaults, and safes.
   ○ C. Do not allow employees to use portable devices.
   ○ D. Encrypt the data.

   Correct answer = D

   Laptop computers, PDAs (Personal Data Assistants), and cell phones can be easily lost or stolen. The best way to protect the information contained within these units is to use encryption. Using a BIOS password is always a good idea for securing access to a system before the operating system loads. However, it is not the most secure method for protecting data. Choices B and C are great ideas. However, they are nonproductive and impractical.

6. **Which type of fire extinguishing agent holds back water in the pipe with a clapper valve, offering time to shut down the system if you happen to get the fire under control?**

   ○ A. Wet pipe.
   ○ B. Dry pipe.
   ○ C. Gas discharge.
   ○ D. Timing pipe.

   Correct answer = B

   With a dry pipe system, water is held far back from the nozzle by a clapper valve. If the system detects fire, there remains significant time to shut down the system if the happen to put the fire out before water is needed. A wet pipe system keeps always keeps water in the pipes that lead to the sprinkler head or nozzle. A gas discharge system doesn't use water. Timing pipe is an invalid selection.

7. **Which type of handheld fire extinguisher should never be used to put out a Class A fire?**

   ○ A. BC.

   ○ B. ABC.

   ○ C. DC.

   ○ D. ACD.

   Correct answer = A

   A Class BC fire extinguisher is rated for chemical and electrical fires. It should never be used to extinguish a Class A rated fire. An ABC extinguisher is rated to put out normal paper or wood burning fires and can be used for a Class A fires. Selections C and D are invalid classes.

8. **Which type of wiring is most secure and least susceptible to interference?**

   ○ A. CAT5.

   ○ B. Coaxial.

   ○ C. Fiber-optic.

   ○ D. High-quality RJ11.

   Correct answer = C

   Fiber-optic cable is very secure and is least susceptible to interference or crosstalk. CAT5 and coaxial cable can be easily tapped and are more susceptible to electrical or magnetic interference. High-quality RJ11 would make a great choice for an analog phone connection but it is irrelevant here.

9. **Which backup type provides the easiest and fastest restore?**

   ○ A. GFS.

   ○ B. Incremental.

   ○ C. Full.

   ○ D. Backup 0.

   Correct answer = A

   The answer to this question is A. A GFS (*Grandfather-Father-Son*) backup strategy using a daily differential backup provides the easiest and fastest restore. With this strategy, the last full and last differential backup tapes are needed to do the restore. All others choices are incorrect. Be ready to handle questions similar to this that contain minimal information on the real exam. Some of the questions on this exam are going to leave you wondering where is the rest of the information.

10. **Which type of backup site will allow you to get your business systems and applications up and running in the least amount of time?**

   ○ A. Hot site.
   ○ B. Cold site.
   ○ C. Warm site.
   ○ D. Fiber-optic site.

   Correct answer = A

   A hot site is considered a site that can provide full business functionality in a very short time. It is the most functional site. A cold site doesn't have any equipment such as servers or workstations on site, and there is usually no active connectivity to external networks in place. A warm site has more functionality than a cold site but would need more equipment and associated components to equal that of a hot site. A fiber-optic site sounds real good; unfortunately, it is an invalid selection.

11. **What should be your first step when creating a DRP (Disaster Recovery Plan)?**

   ○ A. Determine which type of site will be used.
   ○ B. Define the goals the plan is expected to achieve.
   ○ C. Identification of a disaster recovery response team.
   ○ D. Identifying important information to back up.
   ○ E. None of the above.

   Correct answer = B

   Be ready for this type of question on the real exam. The first step needed when creating a comprehensive DRP is defining the goals that the plan will achieve. This will usually include the identification and definition of what is considered to be a disaster or threat to your business. Choices A, C, and D are all items that should be included in the DRP. However, they are not considered the first step in this process.

12. **What RAID level provides the best level of fault tolerance and performance?**

   ○ A. RAID 32.
   ○ B. RAID 0.
   ○ C. RAID 1.
   ○ D. RAID 5.
   ○ E. None of the above.

   Correct answer = D

   RAID level 5 places parity information across all disks in an array. It provides the best combination of fault tolerance and performance of the popular RAID implementations. RAID 32 is an invalid selection. RAID 0 is not fault tolerant. Although RAID 1 or disk mirroring is fault tolerant, it is not considered to provide the best level of protection and performance of the RAID levels.

13. **Which statement best describes a security policy?**

   ○ A. Once documented, it is set in stone and never changes.
   ○ B. Does not typically include an "Acceptable Use" statement or clause.
   ○ C. It is a living document that is never completed or finished.
   ○ D. Should only be followed by security personnel.
   ○ E. None of the above.

   Correct answer = C

   A security policy is never completed. It is a living document that requires continuous updates to reflect changes occurring over the lifetime of a business. Most security policies include an employee Acceptable Use policy statement or clause. A security policy should be provided to all employees, vendors, and third parties involved with the particular business.

14. **What is the activity of sifting through someone else's trash in order to gain confidential information called?**

   ○ A. Rude and lascivious.
   ○ B. Custodial engineering.
   ○ C. Social engineering.

○ D. Dumpster diving.

○ E. None of the above.

Correct answer = D

Dumpster diving is going through someone else's trash with the hopes of finding information such as names, ID's, phone lists, passwords, network information, PINs, account numbers, and other information that can be used for social engineering attacks and access to information systems. All other answers are invalid.

15. **How should confidential company information that is stored on electronic media be handled if it is no longer needed?**

○ A. It should be thrown in the dumpster.

○ B. It should be encrypted and then thrown in the dumpster.

○ C. It should never be thrown away. It is company-confidential.

○ D. It should be given to the security guard.

○ E. None of the above.

Correct answer = E

Any company information that is no longer needed should be disposed of properly. Paper documents should be cross-shredded. Information that is stored on electronic media should be erased. Stay alert when taking the exam. There will be many common-sense questions similar to this that you cannot afford to miss.

16. **What is Two-Factor SSO?**

○ A. Using a user ID and a password to authenticate.

○ B. Using a biometric device to authenticate.

○ C. Using a user ID\password combination and a retina scanner for authentication.

○ D. Using a client/server environment to authenticate.

○ E. None of the above.

Correct answer = C

With Two-factor SSO, a user provides an ID and a password combination and is also required to authenticate with a token or biometric device such as a retina or fingerprint scanner. Using a user ID and a password to authenticate is an example of plain Single Sign-on (SSO). Using a biometric device alone is not considered Two-Factor. Using a client/server environment to authenticate is typically a prerequisite that provides an environment for the implementation of SSO and Two-Factor SSO.

17. **Which of the following are important considerations concerning computer forensics?**

   ○ A. Secure third-party storage.
   ○ B. Following a chain of custody.
   ○ C. Using rubber gloves and sealed containers.
   ○ D. Proper collection and preservation of data for evidence.
   ○ E. All of the above.

   Correct answer = E

   All choices are valid concerning computer forensics.

18. **Which type of risk analysis uses controls such as deterrent, preventive, corrective, and detective?**

   ○ A. Ultimate risk analysis.
   ○ B. Quantitative risk analysis.
   ○ C. Qualitative risk analysis.
   ○ D. Quality control risk analysis.
   ○ E. None of the above.

   Correct answer = C

   Qualitative risk analysis uses deterrent, preventive, corrective, and detective controls to offset risk. With quantitative risk analysis, the risks are guessed and money is appropriated as a means to offset or take care of the aftermath if an event should occur. All other choices are invalid.

19. **What is the formula used to calculate annual financial expected loss?**

   ○ A. ARO + ALE=SLE .
   ○ B. ALE-ARO=SLE.
   ○ C. ALE × SLE=ARO.
   ○ D. SLE × ARO=ALE.
   ○ E. None of the above.

   Correct answer = D

In order to produce an Annualized Loss Expectancy (ALE), you must multiply the Single Loss Expectancy (SLE) by the Annualized Rate of Occurrence (ARO). All other choices are invalid.

20. **According to security roles and responsibilities, whose role includes the responsibilities of backups and restores?**

    ○ A. Manager.
    ○ B. Owner.
    ○ C. Custodian.
    ○ D. User.
    ○ E. None of the above.

    Correct answer = C

    Custodian duties include the verification of system backups and restores as well as keeping updated documentation supporting testing and production efforts. Management is responsible for ensuring that all employees follow security policies and practices and protecting all company assets. Owners' responsibilities include determining the classification level of data, making changes to the classification level, and assigning or delegating who will be responsible for data and the security of data. Users should be responsible and accountable for the data they work with.

21. **Concerning government and commercial data classification, which two data categories describe grave or extreme damage that can result if the public accesses this information?**

    ○ A. Confidential and Top-Secret.
    ○ B. Sensitive and Unclassified.
    ○ C. Secret and Sensitive.
    ○ D. Secret and Sensitive but Unclassified.
    ○ E. None of the above.

    Correct answer = A

    Information that is commercially classified as Confidential might cause extreme damage if made public. Information that is governmentally classified as Top-Secret might cause grave or extreme damage if disclosed. These are the most classified of the two data classification categories. All other choices are invalid.

REFERENCES

http://www.intel.com/support/storageexpress/6736.htm is an Intel Web site that explains the GFS backup strategy in great detail.

http://www.cisco.com/warp/public/126/secpol.html#1a is a Cicso Web site that provides very informative descriptions of network-related security policy practices.

http://www.isc2.org/cgi-bin/content.cgi?category=12 provides a code of ethics for those who are CISSP (Certified Information Security Professionals) certified.

http://www.cert.org/csirts/csirt_faq.html is the very informative Web site that Carnegie Mellon University provides. It details incident-response handling and services.

http://www.nist.gov/ is maintained by The National Institute of Standards and Technology maintains. It has a wealth of information regarding forensics.

http://www.forensics-intl.com/def4.html is the Web site of New Technologies Inc. It has a detailed description of computer forensics.

CHAPTER 7

APPLICATIONS AND SYSTEMS DEVELOPMENT

IN THIS CHAPTER

- Applications
- Databases
- Knowledge-based Systems
- System Development
- Test Tips
- Chapter Summary
- Review Questions
- References

This chapter covers the security-related elements of application/systems development and data storage. We'll discuss the procedures, risks, and safeguards that pertain to database management and security, application controls, and the steps used in system development. As with many of the other topics covered in this book, the concepts discussed in this chapter are as deep as the big blue sea. However, every effort has been made to distill only the facts needed to achieve certification success.

In contrast to earlier chapters, much of what this chapter aims to explain is not deeply rooted in standards. Instead, several of the practices and procedures we'll discuss here provide an outline for systems development. In other words, they offer a framework that organizations can use in their efforts to develop and secure the many systems and applications that come about. Do not be too intimidated by the vagueness of these concepts; the exam is likely to test your fundamental knowledge of the different goals and objectives presented here—not every possible combination of developmental courses of action.

By the end of this chapter, you will have a good understanding of the essential concepts of applications and systems development, with a focus on data warehousing and mining, knowledge-based systems, the systems development life cycle (SDLC), and the different computing environments that applications run in.

APPLICATIONS

The applications involved in *applications* and *systems development* consist of agents, viruses, applets, and other software. It is important to understand the difference between the two basic computing environments and the applications that are associated with each environment. Generally, we speak of applications operating in either a distributed or a nondistributed computing environment.

DISTRIBUTED COMPUTING ENVIRONMENT

A *distributed computing environment* (DCE) is one where the components that make up an application are running across multiple computers in a network. This means that one system could house an application while another contains the database of information that the application refers to. Still a third could handle processing data in the database, making the trio much more efficient than a single system trying to do all these tasks. Indeed, there is no limit to the number of systems that can coexist in a DCE. The players in a DCE could include servers in different geographic locations, running different operating systems on multiple platforms.

The Internet itself is a perfect example of the incredible power of distributed computing. Every time you fire up your browser and key in a Web address, distributed computing is serving you. When you key in that Web site, your computer must query a DNS server for the IP address. Because of the sheer size and dynamic nature of the entire domain name database, it's lucky for us that other computers are handling the storage and maintenance of this information—not to mention the queries for IP address stored within. Because this task is being *distributed* to another, more powerful system, the search only takes milliseconds. Then, when you arrive to the Web site, most of what comes over the wire is simply HTML code; it is your computer that processes the HTML and puts it all together on your screen. Add Java, search engines, streaming media, and the wide variety of other information and applications at your disposal online, and you have a powerful DCE.

Another popular trend in distributed computing is called *application hosting*. Programs can be served up like any other data, providing a platform-independent way for users to connect to the applications they need. This can be accomplished by means of browser-based interfaces, Java applets, or terminal servers. The term client/server is often used to describe this type of remote connection to server-based applications.

Although some of these topics have been covered in Chapter 3, for the exam, you must know which types of applications run in each computing environment. For this reason, we will review a couple of these applications, introduce you to a few more, and categorize them into their respective computing environments. What follows is a listing of the applications native to distributed computing.

Applets

As you know, *applets* are miniature applications that are hosted on a server computer and transferred to a client computer upon request. They can perform a host of different functions and are intended to run within other applications, such as a Web browser. Most modern browsers have the ability to restrict what an applet can and cannot do. Security concerns related to applets include file read/write access, transmission of a user's e-mail address, and reporting about which Web sites a user visits. Although Java Applets and ActiveX controls were discussed earlier in this book, it is important that they be revisited to emphasize their significance in applications development and their place in the distributed computing environment.

 The Internet is a prime example of a large-scale distributed computing environment.

Java

Java is an object-oriented programming language that enables the creation of cross-platform applications. Most browsers contain a component called a Java Virtual Machine (JVM), which enables them to independently run Java-based applets. These applets are light in code and can be run from within browsers on any operating system. Generally speaking, Java applets do not have the ability to read or write files on the client computer, nor are they allowed to make network connections, except back to the host from which they originated.

Another security feature of Java applets is that they are prohibited from starting up programs on the client computer. These inherent security features of Java allow a user to safely execute applets that are not *trusted* (or applets that do not contain a trusted digital signature). In fact, the JVM automatically categorizes any Java applet that is downloaded as unsafe. This is a significant trait because Java applets are often executed without the user's knowledge. However, there are ways to include a digital signature within the code of a Java applet, enabling some of the built-in security restrictions to be loosened, provided one trusts the source. Opposed to the *sandbox* method, which places rigid controls upon the actions of an applet, the implementation of digitally signed applets permits a more flexible operating environment.

ActiveX

ActiveX controls, developed by Microsoft, are very similar to Java applets. Opposed to Java applets, which need to be downloaded every time they are needed, ActiveX controls stay resident. This means that they are downloaded only once and when they are needed again, the system refers to the copy on the hard drive. Take a peek into your Windows folder and you'll probably find some files with the OCX extension. These are ActiveX controls. They can accomplish many of the same tasks as Java applets—and also some that Java can't—but the security concerns are quite different. ActiveX controls can be granted unrestricted access to a computer's resources. For this reason, the only ActiveX control that should be executed is one that is digitally signed from a trusted CA. Thankfully, Microsoft's browser handles this concern automatically.

In Figure 7.1, note that downloading unsigned ActiveX controls is disabled by default. It is possible to change this setting although it is not a recommended action.

Agents

Agents are applications that automatically collect data or perform predefined tasks on behalf of a user. Just like travel agents or real estate agents,

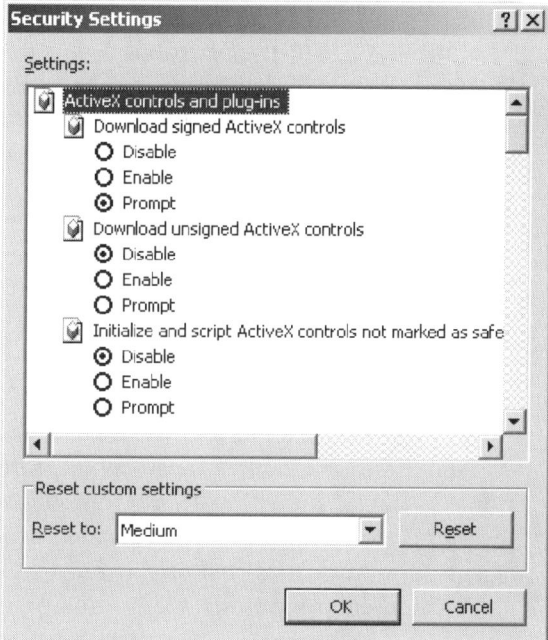

FIGURE 7.1 Internet Explorer security settings.

the digital variety of agents tries to turn complicated tasks into a one-step process for the user.

Agents include a wide range of programs and utilities that share these traits. *Intelligent agents* have the ability to learn and can also possess certain decision-making properties using techniques such as rule-based programming, which will be discussed later. A common example of an agent is one that can browse the Internet in search of data that it has been configured to collect and then deliver it according to a schedule. Popular search engines employ agents (or bots/spiders) that search the Web for new information to include in their databases.

Although they are not the typical example, e-mail clients are another type of agent. Not only can these programs automatically fetch your mail but through the process of filtering, they can also decide the importance of individual pieces of mail. For instance, mail from a particular person might automatically be forwarded to your cell phone upon arrival. Or in another case, mail that contains a certain phrase in the subject line might be deleted upon arrival.

Two common types of agents are known as *static* and *mobile*. While static agents typically stay put, mobile agents are capable of roaming networks, copying themselves to other computers, and gathering all sorts of information. The use of mobile agents raises many security issues since they can be used to deliver viruses, Trojans, and other malicious code. A mobile agent could also send sensitive data back to its place of origin, which might be the lair of a malevolent hacker.

Another security concern that pertains to agents involves the agent itself being attacked. A harmless agent that's going about its business collecting or transporting sensitive data could be attacked by a computer system on the network, resulting in its contents being exposed. This problem is exemplified by the idea of a shopping agent, which travels to several trusted, online establishments looking for the best price. To shop in this way, a user provides the agent (through a browser interface) some form of digital cash, such as a credit card number. When the agent finds the best deal, it makes the purchase. There must be a way to protect these credit card numbers while in transit and encryption techniques are among the solutions being developed to overcome this interesting problem. These kinds of mobile agents, also known as *goal-oriented agents*, are being used in other areas such as mobile computing (PDAs), and online auctions.

 The aim of an agent is to complete tasks for users.

OBJECTS

Objects are self-contained, encapsulated units of code that are divided into a hierarchy of classes and subclasses. A distributed computing environment is an ideal setting for *Object-Oriented Programming (OOP)* in which applications are designed to work with objects instead of actions. This innovative concept in programming is facilitated by a software component known as an *Object Request Broker (ORB)*. When an object is called by an application, the ORB functions as the negotiator between the application and the client. In other words, the ORB provides clients with access to distributed objects. Developed by the Object Management Group (OMG) and embraced by Netscape Communications, the *Common Object Request Broker Architecture (CORBA)* is a popular ORB that is playing a part in the advancement of this technology on the Internet. Related technologies include Microsoft's *Component Object Model (COM)* and *Distributed Component Object Model (DCOM)*.

In general, OOP allows for the speedy development of cross-platform, reusable applications. The modular, recyclable nature of objects also lowers the cost of development because old objects can easily be inserted into new application development projects.

Objects such as Macromedia Flash movies are everywhere on the Web nowadays. If you tinker with Web sites, you might recall seeing an OBJECT tag followed by a CLASSID attribute in some HTML code. This is one method of referring to a distributed object that resides on a remote machine. OBJECT tags allow the browser (the client) to access an encapsulated software component (the object) that is being distributed by the Web site (the server) in the client/server-based, distributed computing environment called the Internet.

LOCAL/NON-DISTRIBUTED COMPUTING ENVIRONMENT

The *non-distributed* (or *local, centralized*) computing environment is one where applications are running in an isolated environment on the local system. This situation presents its own distinct security threats and related safety measures. Understand, non-distributed computing doesn't suggest that you are totally cut off from the world and locked in a closet with no wires attached to your computer. Viruses, for example, can be contracted by way of downloading e-mail, yet they are in this category. Non-distributed defines the environment with regard to the way programs run. The applications in this section are distinguished by the threats they present to the very system they are running on. For more detailed information on viruses, worms, and other malicious code, refer to Chapter 9. The following sections discuss the non-distributed applications that you should commit to memory.

 Security concerns in the non-distributed computing environment are characterized by a threat to the local system.

Viruses

Viruses are small, malicious programs that share common traits such as having bad intentions, running against a user's wishes and without their knowledge, and the ability to replicate themselves. Viruses are also characterized by their synthetic nature, which means that people make viruses—they are not accidents. Delivery techniques include e-mail, infected disks, and HTML code. There are many types of viruses that will be discussed later on. The following list just briefly defines the few that should be remembered within the context of non-distributed computing:

- **Boot-sector viruses:** This brand of virus, also called *Master Boot Record (MBR)* viruses, corrupt or alter information in the primary areas of a disk. They can cause serious problems and destroy an entire disk.
- **Macro viruses:** This type, commonly found in Word and Excel files, are usually considered to be low risk. Some can delete files or add interesting captions to your word-processing documents.
- **File-infecting viruses:** *File infectors* can attach themselves to other programs that appear to be safe only to propagate on the system when the host application is executed.
- **Stealth viruses:** These get their name from the fact that they try to conceal their actions from the user. Masking file size information and removing themselves from memory are two methods by which stealth viruses accomplish this.
- **Polymorphic viruses:** This variety has mutation capabilities that can create a new signature each time. This trait makes them hard to detect because it is a virus's known signature that gives it away.

Worms

Worms are characterized by their ability to operate independently rather than through the use of a host file. They can replicate themselves with the assistance of innocent e-mail servers and unwitting address books, causing rapid spreading. The ILOVEYOU and Melissa worms worked in this way, bringing thousands of e-mail servers to their knees.

Trojan Horses

Trojan horse is a pretty vague term that is attached to any malicious application that masquerades as a harmless one. Trojans can be used to deliver viruses or worms and they can cause quite a bit of damage. Upon installation, the infamous Back Orifice Trojan allows ill-intentioned crackers to gain complete control of a system over a network connection.

Logic Bombs

This kind of nasty, delayed-reaction program (a.k.a. *slag code*) executes itself after a certain event takes place, such as on a predetermined date or at a certain time. There can be any number of harmful actions that logic bombs can inflict on a hapless computer. The original CIH (or Chernobyl) virus worked in this manner. It was programmed to explode on April 26th, the anniversary of the Chernobyl disaster.

APPLICATION HARDENING

Because the applications that power computers and networks are so diverse, numerous, and highly accessible, keeping them all up-to-date and secure is a tall order. To harden an application means to make it more secure by adding features, restricting access, removing unnecessary services, and staying on the cutting edge of software technology. Hardening an application, server, or operating system involves the implementation of many different hardware and software components. User authentication features must be used to their fullest potential on machines with high visibility, such as on the Internet or other network.

Whether it's applying service packs, or configuring firewalls to restrict application access, someone must ensure that applications are kept up-to-the-minute with the latest fixes and the newest (stable) security components. Why is this so important? Glad you asked. The reason is nothing makes a cracker's job easier than finding that they are still able to exploit a security loophole in your system even though it's been widely publicized for weeks or months. You see, the crackers and thugs are reading the news, too. Security advisories (or bulletins) seem to come out daily. They are reporting recently discovered flaws in this application, or that OS, continually. The moment a new security bug is discovered in a particular Web server, for instance, the malicious user will get online and try to take advantage of it.

In order to protect these applications, security professionals cannot slack off in this area. The overwhelming number of security bulletins and associated updates that get released cause many people to ignore the warnings and simply wait for the full product upgrades. This behavior should be avoided—the best defense is to stay current and read, read, read! One way to stay ahead of the game is to subscribe to e-mail security bulletins (or mailing lists) that report updates to the applications you rely on. In the following pages, we will talk about the different types of hardening techniques and how they lower the risk of attack.

 Application hardening involves the use of software mechanisms such as updates and virus scanning as well as hardware-based techniques, such as the use of firewalls.

NOTE

UPDATES

Updates come in many formats and they are the simplest way to bring an application or device up to current standards. Frequent updates can cut support costs because they resolve most technical support issues automatically. This keeps users and administrators happy. Most vendors will supply some sort of update management tool that can automate the process. When dealing with updates across multiple computers in a production environment, it is always a good idea to first test them in a smaller, controlled group.

Many hardware devices, such as routers, include microchips that contain software components. Much like a mini OS, these hardware-based software mechanisms (known as *firmware*) control the functions of the device. Router firmware updates are simplified through the use of the *Trivial File Transfer Protocol (TFTP)*. This protocol is known as trivial because it can be implemented with just a few lines of code, making it ideal for the small chips inside routers. Router manufacturers regularly update their TFTP servers with the latest version of each router's firmware. This process makes it possible to teach old devices new tricks.

If a security bug in a router's program is found, a quick firmware update can usually fix it. To update firmware, typically, a connection is made to the router through a Telnet session where the TFTP server IP address is entered. The firmware update is located, downloaded, applied, and then the router is restarted. As with any other update, it's important to read the manufacturer's instructions before altering your router's firmware. A firmware update could cause problems depending on your network configuration and you might not even need the update. Also, backing up the router's configuration is a good idea before doing an update because could cause data loss.

Another software component that requires regular updates are drivers. Drivers help applications communicate with hardware and a defective driver could present a security problem.

The following list explains the types of updates to be familiar with. These categories apply to the many updates intended for client applications, operating systems, firmware, and the various types of server applications that will be discussed shortly.

- **Hot-fixes:** This type of update takes place while an application is operating. *Hot-fixes* are typically released to address a very specific bug or security concern. When a security bulletin hits the front page, hot-fixes are rushed out to consumers. This rush sometimes produces unreliable—even hazardous—code because little testing is done on these quick remedies. Always read any documentation that accompanies hot-fixes because it just might save a great deal of time. Hot-fixes are usually released between service packs.

- **Service packs:** In Microsoft terms, a *service pack* (SP) is a collection of fixes and updates to an OS or application. They are all-in-one update kits, if you will. Although SPs are mostly just collections of fully tested hot-fixes, they might also include other utilities and updates as well as complete product upgrades.
- **Patches:** Although the terms patch, fix, and update are often used loosely and interchangeably, *patches* have a few features that distinguish them from the others. First, patches are called by this name because they tend to add something to an executable file. They are patched (or inserted) into programs to fix a bug or add functionality. Patches are independent bits of *object code* that integrate with an application that is already running on a system. They can be thought of as bandages because they temporarily fix a bug until a more complete update can be developed.
- **Upgrades:** While updates, patches, fixes, and service packs are normally available as free downloads, product upgrades provide a jump to a completely new version of an application and most often come at a price. The main difference between an update and an upgrade is in the numbers. An *update* will contain minor changes and error corrections, advancing the number to the right of the decimal point in a product version. For example, you might apply an update to ACME Web Server 4.1, bringing it up to Version 4.2. An upgrade, however, signifies a new product release, characterized by an increase in the number to the left of the decimal point. Hence, *upgrading* ACME Web Server 4.2 would bring it to Version 5.0.

HARDENING SERVER APPLICATIONS

The programs that run in the server environment tend to have more responsibilities than their client-based counterparts. Consequently, security bugs or other flaws in these programs will cause more widespread damage, data loss, or privacy leaks. It's also the server apps that are the focus of most attacks. For these reasons, more effort is centered on tempering the applications and devices that provide network/resource access. Let's take a look at how the primary types of server applications are strengthened and why.

Web Servers

The applications that drive Web servers bridge the gap between internal databases and the rest of the world. Machines that serve HTTP requests are among the most widely harassed on the Internet. When Web server applications are installed, the default settings will likely open up a plethora of vulnerabilities, and if your server has not been hardened, you are at risk—period.

Threats to Web servers include Denial of Service (DOS) attacks, breaches in SSL security, and plain old password guessing. Although not widely publicized, there have been attacks on major online vendors that resulted in the disclosure of "private" credit card information. SSL has also had its share of updates regarding security flaws.

Hardening techniques for the high-risk applications that power Web servers include blocking unused TCP/IP ports through the use of firewalls or other means, renaming administrator accounts, adding special software that can recognize (and deny) *spoofed* TCP/IP packets, and employing SSL on all secured pages. Directory browsing should be disabled, along with any unused services, protocols, or executables. If remote management is a necessary feature, make sure that encryption is used and strict policies regarding password strength and rotation are in place. Any other required access points should be fortified with proper authentication techniques. If possible, replace Telnet with SSH. Ongoing installation of the latest updates is also an essential hardening tool. For example, the Nimda and Code Red worms continue to attack Microsoft IIS-based Web servers even though a fix has been available for some time.

E-mail Servers

The first thing that comes to mind when considering the protection of mail servers is viruses. Although many viruses propagate by exploiting the address books of e-mail *clients*, they would never reach the client if they got stopped at the server. Virus protection hardens mail servers in the same way it protects the desktop by placing suspect code/attachments in quarantine. This prevents infections from spreading to the clients served by the system.

Not only can a mail server contract a virus but it can also be brought to its knees by a flood of self-duplicating, infected mail messages that the server itself is immune to. One drawback of mail server virus scanning is a decline in performance because the server must process each message in the search for viruses. Viruses are not the only threat here. The major protocols that mail servers use—SMTP, POP3, and IMAP—have some built-in weaknesses. The *Simple Mail Transfer Protocol (SMTP)* handles client requests for sending mail. An SMTP server should be secured by requiring some form of authentication before processing these requests. This can be accomplished by way of a username and password or by denying requests from outside the subnet/domain. SMTP can also be configured to use SSL to encrypt client/server traffic. If an individual could gain access to the SMTP service, they could bring a mail server down with a barrage of unauthorized outgoing messages.

The *Post Office Protocol version 3 (POP3)* and *Internet Message Access Protocol (IMAP)*, which handle client requests for mail retrieval, typically are secured by a user name and password. However, the standard versions of these protocols are insecure because passwords are sent in plain text and could get sniffed out. Like SMTP, POP3 and IMAP can be configured to use SSL, thus encrypting traffic so that passwords (and therefore mail) do not get stolen. Secure versions of SMTP, POP3 and IMAP are known as SMTPs, POP3s and IMAPs. Kerberos and S/MIME can also be used to secure traffic between mail clients and servers.

Another concern of the e-mail server administrator is *spam*—unsolicited e-mail messages that are often sent to large groups of users at once. The *bona fide* scourge of the Internet, spam clogs servers with huge amounts of messages. Unless some kind of filtering is in place, a mail server will waste everyone's time and money processing these e-mails. There are laws in place (*http://www.spamlaws.com*) governing the transmission of unsolicited electronic messages.

FTP Servers

The *File Transfer Protocol (FTP)* provides a speedy method of transferring files between clients and servers. FTP servers provide user authentication via passwords and they can incorporate password encryption techniques just like mail servers.

One feature of an FTP server that should be disabled in a high-risk environment is the anonymous logon, which allows any user to establish a connection. FTP servers can also fall victim to bounce attacks. This type of attack is enabled by the misuse of the FTP *PORT* command. An FTP server vulnerable to this type of siege can be protected by proper updates or the addition of IP filtering techniques to the server/firewall.

DNS Servers

As discussed in Chapter 5, if DNS (Domain Name Server) zone data is altered, there can be problems. Altering this type of information (known as DNS *poisoning*) can lead to e-mail theft, spoofing attacks, and masquerading Web sites. At the cost of performance, DNSSEC fixes the problem by encrypting zone data. Another set of issues involves the ISC (Internet Software Consortium) DNS server product known as *BIND (Berkeley Internet*

Name Domain), which powers a large portion of name servers on the Internet. Recent reports have publicized several security vulnerabilities in older versions of BIND. These include buffer overflow problems, Denial of Service issues, and exposure to info leaks that can cause the disclosure of sensitive server information by means of specially formatted *inverse queries.*

Inverse (or reverse) DNS queries are requests for domain names when all one has are IP addresses. This is opposite from the typical request a browser sends to a DNS server when you type in a domain name and you get an IP address in return. There are fixes that prohibit the damage a malformatted reverse DNS lookup can cause. As with other types of server applications, staying informed and applying current updates will fix most problems. IP address logging is also essential because in a post-attack investigation, it is helpful to see where IP packets were coming from and going to.

NNTP Servers

Running services on port 119, a Network News Transfer Protocol (NNTP) server provides forums, known as *newsgroups*, that cover a mind-numbing array of topics. The worldwide collection of NNTP servers is called Usenet. Like e-mail servers, NNTP servers are prone to spam and filtering can help blot it out. By nature, access to news servers, usually provided by an ISP, is widely unregulated. This results in an inundation of bogus messages, advertisements, objectionable material, and dangerous code.

Because many NNTP servers allow files to be attached to messages, any type of harmful program could accompany a Usenet message or *post.* For this reason, virus scanning is implemented on news servers, too. NNTP servers are prone to DoS attacks and other insults, so the same general policies as the previous server applications apply to NNTP hardening as well.

DHCP Servers

A *Dynamic Host Configuration Protocol (DHCP)* server is constantly listening for client requests for IP addresses. Malformatted requests can expose vulnerabilities in DHCP servers. One hardening technique for internal DHCP servers that combats this is called *ingress filtering*, where the server or firewall is configured to prevent ingress (or incoming) access to the server from outside the network. If the DHCP server must reside outside the immediate network or beyond the protection of a firewall, it must be prevented from assigning IP addresses to unauthorized clients. If this were to happen, an attacker could gain access to network resources. Reducing the DHCP scope to the smallest workable range can minimize this threat.

Another preventative measure is to extend lease times, reducing the possibility of an unused IP being stolen. Still another trick is to map (or reserve) IP addresses based on the known MAC addresses of authorized clients. There have even been instances where DHCP server vulnerabilities enabled malicious servers to execute harmful commands on the client. The best hardening technique for issues like this—read, update, read, patch, read, hot-fix, and so on.

Data Repositories

A descendent of the data dictionary, *data repositories* are characterized by having large amounts of data from multiple sources. They can store any number of data types but they mostly contain information related to the internal structure of an organization, such as objects, business models, or even references to other databases. The next section in this chapter covers databases and database security in more detail.

The use of directory services is one method used to harden a data repository. The X.500 standard specifies a structure for this type of global, centralized directory. The familiar Lightweight Directory Access Protocol (LDAP) provides directory services. While offering much of the functionality of a full-blown, X.500-defined OSI directory service, LDAP does not require as much computing power. Other products such as Microsoft's ADS (Active Directory Services) or Novell's NDS (Novell Directory Services) offer directory services for their respective environments. ADS and NDS act as the central network authority, providing access to virtually anything in a distributed computing environment. Printers, files, programs, databases, and other resources are classified in the directory, and access is given to users based on their predetermined rights.

File/Print Servers

The most important aspect of hardening an intranet file/print server is securing the physical machine itself. Other than issues regarding unauthorized physical access, the same threats apply to these types of servers as with the others. Most issues are managed with the proper authentication techniques, virus scanning, and so on. On Internet-connected Intranet servers, unauthorized access from outside the LAN must be prevented in order to protect resources. Firewall software or hardware can block most traffic from the Internet easily. If authorized connections must be made to intranet servers from the outside world, they should be secured by VPN, SSH, or another sound method.

REMEMBER THE CIA TRIAD?

Illustrated in Figure 7.2, Confidentiality, Integrity, and Availability are three terms that arise often in the discussion of security. Within the context of application hardening, they are at center stage. Regardless of the type of server application you are hardening, several practices that address this trio of concerns should always be observed. To harden the confidentiality of server data, encryption-enabled protocols like SSL (aided by digital certificates) and SSH should be used. Strict authentication measures are also essential.

FIGURE 7.2   The CIA Triad.

To harden the integrity of server resources, elements such DNSSEC with its hashing capabilities should be added to the mix. Because viruses and many other attacks can affect the availability server resources adversely, virus scanning and packet/port filtering must be put into place. Data backups, secondary servers, RAID arrays, server clusters, mirroring, replication, backup power sources, and other redundancy systems protect availability as well. In general, the avoidance of so-called *single points of failure* helps to protect resource availability.

Logging application activity is one way to help detect attacks that happen unexpectedly. Detailed logs can also help prevent future attacks. However, logging itself won't do any good unless someone reads the logs.

Always question the default installation settings and paths of any server application. If everything is not where it's supposed to be, an attacker might give up before finding application-critical executables or databases.

Have your server page or beep you when it experiences a serious error. Also, remember that password-based access methods are susceptible to brute force password guessing, so tighten password policies to prevent access. Again, it is essential to stay on top of updates via mailing lists or security bulletins. Usenet newsgroups also provide a valuable source of helpful information concerning bugs and fixes. Physical security procedures apply to any server and the applications it's running because a potential threat could be within the walls of your organization.

 Hardening methods that apply to all types of servers include updates, logging, encryption, strict authentication, and disabling unused services and user accounts.

However clichéd, memory aids such as the CIA triad can be very useful. So much so, that many successful test takers use memory-jogging devices like this—known as *mnemonics*—to help burn all kinds of data into their brains. Do you need proof? Recall the names of the Great Lakes. You probably think of the H-O-M-E-S mnemonic right away. Or maybe, for all you music fans out there, you once used the mnemonic, *Every Good Boy Does Fine*, to remember the notes in a musical staff. Not only can these help you remember groupings of items, but they will also assist in recalling the specific order of items in a group. So go ahead, make up a few of your own. Put them to a tune. Be creative. You'll be surprised how much they help on exam day.

DATABASES

The subject of databases is expansive, to say the least. Fortunately, this section only contains the concepts, terms, and security issues that you'll need to understand for the Security+ exam.

In short, any collection of information that is grouped together in a structured manner can be called a *database.* The phonebook on your shelf is a database. However, when information is added to an electronic database and cross-referenced, it turns into a powerful and dynamic resource. Electronic databases enable users to store, access, modify, and retrieve data while enjoying total control over which elements of the database they wish to work with. When you look up a number in the phonebook, you are limited to searching by last name but what if you're not sure how to spell it? The only option is to try every different spelling and waste valuable time. In a computer database of cross-referenced names, addresses, and phone numbers, the process can be reduced to seconds. The more tidbits you know about someone's entry in the database—such as their first name or zip code—the quicker it can be found.

When discussing the popular relational database model, some traditional database terms are often used synonymously with their modern equivalents. A *relation* is simply a table of rows and columns in a relational database. Refer to Table 7.1 to avoid confusion in this matter. You might see these terms substituted for each other, so it's a good idea to know them up, down, and backwards.

TABLE 7.1   Interchangeable Database Terms

Traditional Terms	*Standard Terms*	*Relational Database Terms*
File	Table	Relation
Record	Row	Tuple
Field	Column	Attribute

What follows is a brief list of the terms commonly used when discussing databases. This is not meant to be a *Databases 101* lesson—just a small word list to help you understand the concepts in this section.

- **Attribute:** The component that describes an object or entity in a database. In a relation, attributes are represented as vertical, named *columns.*
- **Cardinality:** The number of *rows* (or *tuples*) in a relation. Keep in mind that cardinals are red so you will remember the three R's of cardinality: Red, Row, Relation.
- **Cell:** The point at which a row and column intersect.
- **Degree:** The number of attributes in a relation.

- **Domain:** The set of all possible values that an attribute can have.
- **Filter:** Criteria used to include or exclude specific information in the results of a database query. Also used to sort results and filter database input.
- **Index:** Typically used in large relations, indexing deals with the structure and reordering of tuples in a database.
- **Metadata:** Data that describes a database, or simply *data* about *data*. Contains relation and attribute names that form the structure of a database, information about relationships, and data on who's allowed to access what in the database.
- **Query:** This is a command used to fetch and display requested information from a database.
- **Relational keys:** These are used to establish relationships between relations. *Primary keys* uniquely identify each tuple in a relation. Each relation contains only one primary key. *Foreign keys* are used to identify tuples from a separate relation, thus enabling relationships to be made across different relations in the database.
- **Relationship:** This specifies a link between tables (or relations) in a database. Different database models establish relationships in different ways. Relational databases make relationships by connecting tuples between multiple relations. There are three types of relationships to be concerned with. A *one-to-one relationship* is where *one* tuple in relation A is related to *one* tuple in relation B. A *one-to-many relationship* is where one tuple in relation A is related to one or more tuples in relation B but the individual tuples in relation B are restricted to one-to-one relationships with tuples in relation A. Finally, *many-to-many relationships* describe the situation where both relations contain relationships to multiple tuples in the other's relation. Dr. Seuss, anyone? Many-to-many causes problems with data redundancy, making modifications, updates, and deletions more complicated.
- **Tuple:** In a relation, tuples are represented as named rows. They are essentially horizontal rows of attributes. They are collections of facts pertaining to single entities or objects.
- **View:** This is a method of manipulating and controlling the display of (and access to) data. Views are used to control who can change or view individual elements of a database. For instance, an administrator might use a view to prohibit social security numbers

from being displayed when queries are made to a personnel database. A view can control access to data based specific user accounts, or on groups of users.

TYPES OF DATABASES

The most significant types of databases are listed below. Most databases encountered in the real world are relational but each has its own characteristics that need to be explained.

Flat Databases

This variety is the simplest to program and contains the least amount of structure. A single flat file comprised of one table can be the source of data in a flat database. A comma-separated file of names, e-mail addresses, and numbers is an example of a simple flat database. This works fine with a small collection of information. Problems with this kind of database come to light when a flat file starts to grow larger. Flat files generally waste space as a result of their structure, so queries can take much longer as the file gains weight. In addition, they are not suited for complex queries since a program must make separate searches through the data, then sort through everything to deliver results. You might recall from history classes that people once believed the world was flat. Flat databases share a similar flat view of the database world, so they are limited to simple tasks. The flexibility of other types of databases solves these problems.

Hierarchical Databases

Similar to the file management technique used on PCs, *hierarchical databases* contain an upside-down tree where one table resides at the top and others branch out from beneath it. Connected by links known as *pointers*, so-called *parent* and *child* tables form the relationships in this kind of database. This is a good memory aid as parent tables can have many children but a child table can only have one parent. This model is an improvement over flat databases but there are still problems with data redundancy. One-to-many relationships work well with the hierarchical model.

Relational Databases

As we discussed, the concept of *relations* is the driving force of this popular database model. The actual placement of relations in this type of database is insignificant because they are referenced by name and connected by dynamic relationships. Opposed the hierarchical model, a user need not know the structure of a relational database to access or modify it. This model makes efficient use of the space it's allocated because redundancy is kept to a minimum. It also enables complex queries that take full advantage of the inter-connected tuples and relations. Remember that tuples and relations are synonymous with rows and tables. With the assistance of detailed metadata, relational databases offer outstanding database administration utilities.

Object-oriented Databases

Developed to blend nicely with object-oriented programming languages such as Java and C++, *object-oriented database (ODB)* technology offers advantages for programmers. Because the database programming is done in an object-oriented language native to the database itself, it does not require translation and you end up writing less code. Another benefit of this model is a reduction in maintenance. The systems used to manage ODBs are called *object database management systems (ODBMSes)*. ODBs can operate at higher speeds than relational databases, but they are not as widely used. Some technologies can incorporate object and relational databases into a unified system, called an *object-relational database*. ODB's use *data models* that classify objects and give them unique properties. They also support encapsulation and the use of *object identifiers (OIDs)*. One popular method for establishing relationships in an ODB is called *inheritance*. This is when objects of one subclass inherit properties from objects of a higher class (or superclass).

Network Databases

Lending itself well to many-to-many relationships, the network database model relates data groups to each via named *sets*. Similar to hierarchical databases, network databases support the parent-child model but children in network databases can have more than one parent.

 Different database models establish relationships in different manners. Relational databases use relations. Hierarchical and network databases use hierarchies of parents and children. In object-oriented databases, objects inherit classes from higher-level objects.

DATA WAREHOUSING

Before data can be mined, which we'll learn about shortly, it must be collected in one place. A *data warehouse* is a central storage area for relevant business, production, and customer data. Supplied by a mainframe server, warehouses generally contain historical information pertaining to the business activities of an organization. Because a data warehouse is constructed for the purpose of finding trends, by design, they facilitate data analysis. In order to keep a warehouse up-to-date, its contents must be maintained (or *scrubbed*) to remove irrelevant and outdated information.

The term *data mart* is often heard in the discussion of data warehousing. To avoid mixing the two up, keep in mind that a data mart, which can be derived from a data warehouse, is where metadata is stored. A data warehouse is where the actual databases are stored and does not contain metadata.

Another similar term to look out for is *data dictionary*. Comprised of database details and other management or accounting information, data dictionaries are set aside for use by system developers and are typically concealed to prohibit user intervention. The data dictionary is also the predecessor of the *data encyclopedia*, which contains even more information and is used in data modeling procedures.

 Data warehouses supply data mining applications with a stockpile of historical information.

DATA MINING

While warehousing stores data, *data mining* aims to make sense of this data by finding the hidden correlations buried within. Data mining techniques facilitate the analysis of large amounts of information—such as historical business data—in order to extract patterns, relationships, and trends. Although data mining is being put to use in many fields, it's enjoying widespread use in the e-commerce arena. The idea is that through the use of specialized algorithms, mining applications can sort through heaps of data and uncover connections that are invisible to a human analyst. This can be extremely useful to online retailers, for instance, by determining their customers' purchasing patterns.

Through its complex procedures, data mining programs can quickly answer specific questions such as, Who is likely to buy more expensive items and at what time of year? What purchases can be expected next year from buyers of coffee machines in Rhode Island? Through analysis of sales history and comparison of all the related data, these and other more intricate questions can be answered. The Holy Grail of data mining is not as easy to obtain as it is to explain. Specialized mining techniques require quite a bit of planning and work. One must first have a clear, concise idea of which particular pattern or trend needs to be uncovered. Numerous companies are devising ways to offer custom-tailored data mining services to other businesses as this process gains popularity. Data mining is expected to grow by leaps and bounds in the coming years, which will inevitably result in more effective applications that are easier to implement. The components of the data mining process that you should know well are listed shortly. Mining attempts to expose hidden information in data such as the following:

- **Associations:** This involves the effort to uncover events that are interconnected, such as how often computer buyers include a new monitor with their purchase.
- **Classifications:** This involves the search for new patterns in the chaos that result in a restructuring of the existing data.
- **Clusters:** This involves the search for previously unknown sets of facts. Cluster data is usually accompanied by visual representations that display these uncovered groups of facts.
- **Forecasts:** This involves the search for recurrences in data that can aid in making predictions about future events.
- **Sequences:** This involves the hunt for instances of events that are followed up by another event after a period of time. For example, instances of customers buying a certain model of motorcycle, and then returning a month later to purchase a more comfortable seat. The discovery of this sequence could tell a company that the seat in question needs to be redesigned.

Another key phrase to look out for is *Customer Relationship Management (CRM)*. Data mining is one of the tools used within CRM. The goal of a CRM project is to learn about the behavior of customers in order to provide higher quality customer service, more efficient sales techniques, and ultimately, to increase revenues.

Web mining, another tool used in CRM, is an e-variation of the standard data mining process, with provisions for data collected via a Web site. These provisions include content, usage, and structure mining. Content mining works with data from Web spiders and search engines. Usage mining analyzes data related to a user's environment—such as the browser, and so on—as well as information collected by way of user submitted forms. Finally, structure mining analyzes data collected from user interaction with the URLs in a Web site. As with traditional data mining, Web mining utilizes custom applications that don't adhere to any strict standards.

DATABASE SECURITY ISSUES

Just like access to any other server resource, the right to view or modify data in a database is mostly controlled through the use of user accounts, roles, and security classifications. Therefore, the same policies regarding passwords and other user authentication techniques are observed when dealing with database access.

The *database administrator (DBA)* is the individual who has total control over all aspects of a database. Other, more restrictive administrator accounts can be created that allow trusted users to perform lower-level tasks such as creating other user accounts. A DBA can also create *roles* that have specific security permissions. For example, a role called Developer could be defined for users that engage in development tasks. The Developer role would allow users to do things like create tables in a database. This is a risky permission to give to the wrong person, so the right people must be selected carefully. Database security issues present unique challenges and the following sections will explain the major risks as well as the preventative measures used to keep things secure.

AGGREGATION

In the context of databases, *aggregation* is a condition where one has an assortment of low-level access rights, which, when combined, grant access to information of a higher sensitivity. In other words, the security level required to access the highly sensitive stuff is greater than the security level required for the low-level data but a collection of low-level rights gives you access to the sensitive data nevertheless.

The problems related to aggregation have to do with classifications. Careful thought must go into classifying aggregates in a database. For instance, if an aggregate is classified as secret, but the separate elements needed to access the aggregate are unclassified, you might run into trouble. In this example, an unclassified user may be able to inadvertently access the aggregate.

In another example, imagine that individual records containing the phone number of a person in a group are unclassified but the record that combines *all* of the phone numbers of *everyone* in the group is classified as secret. The secret record of the combined numbers is the aggregate in this example. So, if you were an unclassified user, you could obtain the numbers one-by-one and effectively put the rest together yourself. Therefore, through the process of aggregation, you've obtained some data classified as high-level even though your user account has not been granted high-level access. It's not as tricky as it sounds and a test question about aggregation will probably give itself away in the wording. Just be careful that the terms aren't reversed on you and you'll be okay. Study the differences between aggregation and inference, as they are all similar. These similarities will be used to throw you off course.

INFERENCE

Literally, to *infer* information means to gather it by some means, such as deductive reasoning or through some amount of previously known information. The word, *inference*, is synonymous with the words conclusion and deduction. Inference, as it pertains to databases, this book, and the Security+ exam, however, has a more profound meaning. So, for our purposes, *inference* is a problem where unauthorized users deduce (or infer) data at a sensitivity level that they haven't been granted rights to access. It is commonly said that the security problem here is in *prohibiting the inference of unauthorized data*. This sounds similar to aggregation but these explanations should be all that is needed to make a distinction. Questions about inference might mention low-level data *portraying* high-level data. Inference attacks try to take advantage of the inference problem. One preventative measure used to counter inference attacks is called *cell suppression*, which conceals the cells in a database that might enable an attack. Another method used to combat breaches in security caused by the inference problem is called *polyinstantiation*.

POLYINSTANTIATION

Here's one that's not in your dictionary. The use of *polyinstantiation* enables multiple copies of the same data to exist in multiple places in a database, each having a different security level. This security model enables multiple levels of authorization and protects against inference attacks. In database lingo, polyinstantiation is when a *relation* has multiple *tuples* with the same *primary keys*, all stored in the same database. The defining factor is that the primary keys in this setting each have their own security level. Polyinstantiation works by tricking low-level users into thinking they're viewing high-level data. Because polyinstantiation promotes the storage of information in more than one area of a database, every instance of duplicated data must be updated when there is a change. Without simultaneous updates, there is a risk of data integrity loss.

Another process that proposes a risk to data integrity is known as *denormalization*. A database is denormalized to increase its processing speed, but the duplication it introduces to the database causes the same problem as polyinstantiation.

PERTURBATION

To *perturb* means to agitate or displace. *Perturbation* is another tool used to fight inference attacks; it seeks to confuse intruders. This is accomplished by infusing phony information into a database in the hope that an attacker will become frustrated enough to give up and go elsewhere. Perturbation is also called *noise*.

PARTITIONING

Partitioning is used to split a database into multiple parts. In the context of security, the idea is to complicate the inference of information by unauthorized users. If the interconnected parts of a database are difficult to locate, the unauthorized discovery of sensitive information turns into hard work.

MULTILEVEL SECURITY

Multilevel security (MLS) is a system that employs a centralized structure of security classifications called *labels*, which are applied to users and objects. The MLS model makes use of mandatory access controls (discussed in Chapter 2) to manage database access. The component that defines these security labels in an MLS is called a *security policy*.

An MLS system is characterized by its ability to allow multiple classification levels of data to be simultaneously executed. The biggest threat to security in the MLS model is the inadvertent opening of *covert channels*. Many MLS-based systems are susceptible to this method of sidestepping the built-in security and solutions to the covert channel problem are notoriously difficult to implement. In an MLS-based database, a *covert channel* is an exposed link between high-classification data and low-classification data. In effect, if someone can circumvent the access controls in the MLS system and find a way to leak information from high to low levels, they have tuned in to a covert channel.

In contrast to the multilevel secure approach to protecting data, the *system high* model provides a more restrictive and less efficient method of access control. To access a database under the system high model, a user (or any other connected computer) must have clearance to the highest level of data contained in the system. For example, if there's some top-secret info in the database, a user must have top-secret clearance to access the database.

DATABASE MANAGEMENT SYSTEM (DBMS)

This is where working with databases starts to resemble actual fun. A *database management system (DBMS)* is an application (or suite of applications) that creates, controls, manages, and provides access (through the use of *queries*) to data in a database. There are many players in this field and the competition is quite fierce.

A DBMS can be based on several different query languages but SQL (Structured Query Language) is one of the most popular languages used. SQL is the quasi-standard query language for use with relational databases although different companies add their own extensions to SQL, thus creating proprietary versions of this supposed standard. SQL is pronounced either like the word, *sequel*, or spoken as *S-Q-L*.

Although the folks at Microsoft might have put SQL on the map with their SQL Server suite, this query language was originally developed in the mid-1970s by IBM and is currently incorporated into the products of other companies such as Oracle and Sybase. ANSI standardized an early version of SQL in 1986, which was updated in the early nineties and is now called *SAG SQL*, an acronym for *SQL Access Group*.

Whichever query language is being used, they all share a common goal. They offer a means of interacting with a database through the use of various commands and provide rules that govern how database queries are constructed. In turn, the DBMS executes those commands and delivers the goods. DBMSes also have the ability to print reports from the results of a database query and some can enhance these outputs with charts or graphs.

What follows is a simple example of a DBMS at work processing a query. Let's say a user asks a database to return any records it has containing the location of post offices within 20 miles of their home that are accessible 24 hours a day. The DBMS references the database using specific commands native to the query language it's running. The query can either be handled by a graphical interface that converts user input into the right commands or by the user typing the commands directly into the system. Either way, the DBMS will search the database, find any results, and return them to the user. An application can also be programmed to return a custom response if no matching data is found. Additional features can enable a DBMS to deliver information related to the query, although it wasn't specifically asked for.

The extensive use of relational databases has led to the development of DBMSes that are specifically designed to manage them. These DBMSes are called *Relational Database Management Systems (RDBMSes)*. Much advancement is being made in this field and more complex query languages are currently being developed. DBMS architecture that supports these newer, fourth generation languages are sometimes called *4GL systems*.

 SQL is a query language standard used by the database management systems of several manufacturers.

GARBAGE COLLECTION

Now this is a straightforward term. Also known as *storage reclamation*, garbage collecting is a method used to free up space in a database. In a typical database where objects are added and not overwritten, database size can become a problem. Usually implemented in an automatic process—for example with a Java program—garbage collecting applications check for objects that aren't referenced by the database anymore and destroys them. These programs perform housekeeping duties for the database administrator because the admin has better things to do. This one is painless to learn—garbage collection recovers space.

ONLINE TRANSACTION PROCESSING (OLTP)

Databases that deal with customer interaction must operate in real time. Typically distributed over multiple computers in a network, *Online Transaction Processing (OLTP)* mechanisms provide immediate response to requests. In an OLTP system, a process called *brokering* facilitates network distribution. In this type of processing, every user request is called a *transaction* and real-time results are what an OLTP system delivers. Automated teller machine transactions are a perfect illustration of OLTP. Among others, the airline industry uses OLTP systems.

Transaction controls exist that protect the integrity of transactions. Transaction control objectives provide quality assurance with regard to incoming and outgoing data in a system. *Input controls* check user entries such as date/time and ensure that other types of input are complete and correct before they are processed. *Output controls* handle the accuracy of data that a system delivers, providing a checkpoint for information before it's printed or displayed. In a similar way, *processing controls* search for inconsistencies within data that is in the midst of input and output.

KNOWLEDGE-BASED SYSTEMS

A *Knowledge-based System (KBS)* is a system that utilizes human knowledge stored in a knowledge base to solve problems. Because a KBS seeks to solve problems that typically require the expertise and reasoning of a human being, the process is not guaranteed to work. Applications that work with information in this way are said to be *heuristic*, as opposed to *algorithmic*. A knowledge base will contain whatever type of information the programmers need it to process but a few specific categories include rules, facts, attributes, relationships, and events. The two types of knowledge-based systems you need to be familiar with are expert systems and neural networks.

EXPERT SYSTEMS

Although the terms KBS and expert systems are often used interchangeably, for our purposes, we will make a distinction between the two. An *expert system* is a program that solves problems by emulating human reasoning. Said another way, an expert system tries to reason like a person, using a database of knowledge gathered by experts in a particular field. Standing between the user and the knowledge base in an expert system is the *inference engine*—the component that actually gets its hands dirty. The user makes a query through an interface; the inference engine analyzes the query and then consults the knowledge base. Once the system comes to a conclusion, it will deliver its results back to the user. A few fields that benefit from the use of expert systems are medicine, finance, and the oil industry. Oil companies will use them to assist with drilling issues and questions regarding geology, while doctors use them for help diagnosing certain diseases. Financial organizations use expert systems to make forecasts about future trends in the markets.

There are different ways to develop expert systems. These programming models are known as *paradigms*. The most common method is referred to as *rule-based programming*. Using the familiar If and Then parameters, rule-based programming employs preprogrammed *rules of thumb* to provide answers for any given situation. Each individual rule contains an If parameter and a Then parameter.

A more complex system may also include And/Or operators. The If parameter tries to compare user input to patterns in the data. This is where the inference engine does its work. In a process called *pattern matching*, the engine will attempt to match the facts provided by a user with established patterns in the database. When a match is found, the inference engine determines the rule or rules that apply. Once a rule has been singled out, a course of action is determined by the Then parameter. It contains a group of actions that pertain to each rule. One by one, the applicable rules are processed and the recommended courses of action are delivered to the user. Expert systems are fed rules and actions; they do not learn with time. They simply contain programmed responses to recognizable queries.

In the following example, we will look at a simplified version of rule-based programming at work. You aren't feeling well, so you consult a medical expert system on your PC. You tell the interface that you have a cough and a runny nose. The inference engine examines the data and looks for a pattern containing the Ifs: Cough and Runny Nose. A match is made. The rule Cold is determined to be applicable to the situation. Finally, the Then parameter of Cold is referenced, which contains the action to be taken. The inference engine collects this action and delivers it to your screen—Eat Chicken Soup.

Another example of a supercharged expert system is IBM's Deep Blue chess playing computer. In a rematch that ended after six games on May 11, 1997, Deep Blue defeated world champion Garry Kasparov at his own game. This landmark event in computing (and chess) was made possible by IBM's tireless research team. They entered the moves from hundreds of famous chess matches going back 100 years and more. The computer had so much data at its disposal, and could evaluate such a large amount (2,000 per second) of chess positions; the champ didn't stand a chance. The assumption is that the Deep Blue computer that beat Kasparov was a "learning" system but it didn't really employ artificial intelligence. Alternatively, Blue used a rule-based expert system that placed a value on each chess piece, and then decided between thousands of possible moves based on those values. It was determined that neural network systems are great at many things, but trying to think like a chess champ is not one of them.

NEURAL NETWORKS

A *neural network* is a more ambitious method of processing data that attempts to actually think like a person. While expert systems emulate human reasoning with static rules and predetermined actions, neural networks operate in a fashion similar to the human brain. In a manner of speaking, they simulate the nervous system of biological organisms, like us! This is accomplished by approximating the physical structure of the networks of neurons our brains use to solve problems. This is what is known as *artificial intelligence (AI)*. We won't get into the complex programming that is necessary to achieve this but suffice it to say that these systems aren't developed overnight. Not only can neural networks solve problems that computers aren't historically good at solving, but they can also learn new things. Once a neural net is configured (or *trained*), over time, it can make adjustments to the interconnections of "neurons" in its "brain." In other words, they can learn from examples, successes, and failures.

Another reason these systems are called networks is that they are typically comprised of multiple processors in many computers, working in parallel unison, with each layered neuron processing its own bits of data. The more processors that are interconnected, the higher the rate of learning will be and the more intuitive the system will become.

 Expert systems are typically rule based and do not learn over time. Neural networks can learn over time by comparing successes and failures.

SYSTEM DEVELOPMENT

Poorly designed software and hardware components cost companies and users millions of dollars every year. Although it's difficult to produce systems without bugs, and people will always try to utilize software/hardware in ways that the designers never dreamed of, managed methods of development exist that can minimize the inherent bugginess of the systems we rely on. This results in more flexible, secure, and functional applications and devices.

Thoughtful planning, careful selection of team members, orderly division of responsibilities, extensive documentation, and adherence to standards are the ingredients for successful system development. Whatever the reason—user demand, new technology, or the endless drive to stay competitive—organizations will continue to seek ways to develop better, more secure technology. This section aims to explain the processes, security controls, and standards that pertain to the system development process.

SYSTEM DEVELOPMENT LIFE CYCLE (SDLC)

The System Development Life Cycle (SDLC) is a structured, multitiered approach to developing, implementing, and maintaining *Computer Information Systems (CISes)*. It functions as a road map for the teams of software engineers, system analysts, and other individuals involved in the research and development of new systems or those under modification. Rather than being governed by standards, the SDLC method can be customized to fit the needs of an organization. Its goal is to break down the different steps of development and then divide those steps into smaller tasks that are completed along the way.

The SDLC promotes an ordered flow of progress as the accomplishments of one step in the process are applied to the subsequent steps. For this reason, the terms *waterfall* and *ladder* are used when describing SDLC methodology. However, the basic waterfall approach assumes that each step is completed correctly along the way. This is just like a real waterfall where the water only gets one chance to fall. This inherent weakness of so-called *linear* progress has been corrected by the *fountain* approach to systems development, where amendments to the underlying SDLC architecture allow for steps to be reiterated. In other words, certain developmental phases can be cycled through repeatedly until the collective achievements suggest that the step in question is complete and the team can move on to the next step with confidence.

 SDLC promotes broad documentation during all phases.

The successful execution of the SDLC process is dependent upon the people who manage its implementation as well as the people who complete each specific task. It's about teamwork. The right people need to be selected for the right jobs. The process will also move along more efficiently if individual team members can see that their work is actually advancing the project. One aspect of SDLC that helps with morale issues, as well as overall performance, is its extensive use of documentation. The initial goals and outlines, as well as the progress along the way, should be documented as fully as possible.

The security concerns related to the SDLC must be addressed continually throughout the process. Concerns include separation of duties, protection

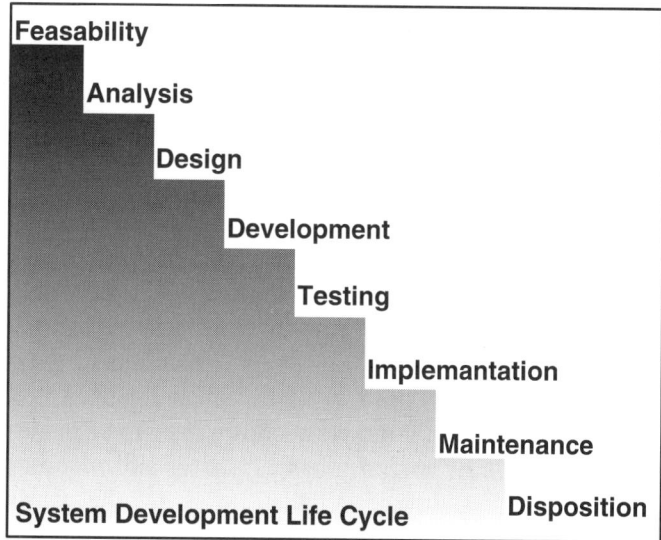

FIGURE 7.3   Steps of the SDLC process.

of programming code in production, and certification/accreditation procedures.

Although the steps in the SDLC process can vary greatly, they generally follow an outline that includes the steps shown in Figure 7.3.

Feasibility

Once the possible need for the project has been identified, this first step aims to answer the initial questions one might expect to hear, such as, What will the project cost? Does the project pose any security threats? Is it legally and economically feasible? What are the risks? Will it have the support and sponsorship of those higher in command? Who will the team be comprised of? These questions and more are answered and then documented. The results of the feasibility phase are used to determine whether or not the project will get the green light.

Analysis

If the go-ahead is given, the *analysis* phase is set into motion. It aims to define and document further, such as security concerns, user requirements, and system performance. During this phase, alternatives such as purchasing an existing software package (as opposed to creating a new one) are evaluated. Decisions regarding outsourcing some of the workload are made at this step as well. After these concerns are evaluated and documented, the results are proposed to the organizational managers, who might choose to develop a product from scratch, purchase and modify an existing application, or scrap the project altogether.

Design

During the *design* phase, the system is reviewed and designed to meet the requirements laid out by the analysis phase. Plans are developed to convert existing data to the new system. Decisions are made regarding which operating environment to use, system architecture, and input/output. As stated before, the steps in the SDLC can be repeated as necessary. However, in the design phase, concrete decisions need to be made that usually prohibit a return to the analysis phase. It marks a firm commitment and a real investment in the project. This is not to say that a project has never passed the design phase only to be scrapped later; it just gets expensive at this point. This phase is crucial and proper documentation is key in the success of the design phase as well as system development as a whole.

Development

In the *development* phase, the results of the design phase are transformed into a comprehensive information system. This phase primarily focuses on the applications that drive the system but it also deals with the hardware and communications requirements of the project. Hardware related to the system will be assembled and hardened. User manuals and other documentation begin to take shape at this stage. *System Security Plans (SSPs)* and *Contingency Plans (CPs)* are also developed here. Other operational requirements are worked out, such as future training techniques for the new system. In addition to documentation and other security-related concerns, the development phase mainly aims to get everything in line for the next phase: testing the system.

Testing

In accordance with the SDLC policy of extensive documentation, developers don't enter the *testing* phase until a clear *test plan* has been written. The plan usually outlines the desired results of testing and defines the time constraints placed upon the phase. Once the modular software components that comprise a system are ready for a trial run, they are tested both individually, and then as a whole.

The interdependencies between software modules are tested to see if they will actually assimilate as planned. The operating system (OS)-specific questions are answered, such as, Will the system run on the platform it was designed for? It is during the testing phase that a beta version will be released to a controlled group in order to collect reports about strong points, weaknesses and bugs. User participation in this phase is referred to as *acceptance testing*. It is often difficult for developers to step outside themselves and give objective feedback from a user's point of view, so this aspect of testing is highly valued. Questions regarding the application's performance are answered at this point. This phase also tests the security related areas of the project, such as access control and data security. A project might see many reiterations of this phase before entering implementation, where things really start to come together.

Implementation

This phase is entered once the system testing is complete and the user acceptance has been established. It is during *implementation* that the system modules are translated into programming code and integrated with any hardware components that the system specifies. Exhaustive code reviews in this phase attempt to weed out previously undiscovered bugs and security concerns. Data conversion takes place to update the format of information or files to meet current system requirements. The system will then be put onto a production machine and evaluated.

Any certification or accreditation requirements are addressed and developers ensure that the system meets any regulations or legal guidelines that were imposed on the project. Standards are addressed and the training plans defined in the development phase are carried out. This phase leads to final production of the system, so it will not be labeled complete until it is functioning properly in a production environment.

Maintenance

Because the *maintenance* (or *operational*) phase will continue throughout the life of the system, a substantial portion of project funding—sometimes more than half—is reserved for system maintenance. Although maintenance is the phase that never ends, it is what keeps developers happily employed for years. Once users rely on an application, they will have an ongoing need for updates, upgrades, and product support. Over time, new users will be introduced to the system, so the need for training will never go away.

An application will also have to adapt to newer, faster hardware. Improvements will be made and features will be added. As a result of standard security audits carried out in this phase, security leaks could surface, requiring patches to be written and distributed in a timely fashion. Bugs will be discovered by users and the maintenance team alike and will need to be fixed. Another application or a new networking requirement could be thrown into the mix and the system might need adjustments to make it play nice with these components. People will want new functionality or add-ons. An application might need to be ported to an operating system for which compatibility was not originally planned.

A team must also coordinate its efforts to document the many modifications that are made and publish new manuals for each version of the application. If a new requirement is identified that entails major modifications of the system, the whole project might indeed revert to the beginning phase.

As you can see, this is quite a bit of work. A large portion of the maintenance process is the continual evaluation of a system to make sure it is meeting the needs originally defined in the early phases, or meeting the redefined needs of a system that's been overhauled. It is an expensive, unending quality-control undertaking. If you have been computing for some time, just think back to an early version of an application that you have relied on for years. Recall updates to your favorite programs that provided accessibility to the many new features that have come about in networking, cryptography, multimedia, hardware, and operating systems. If you have been affected by these advancements, you've witnessed the maintenance phase of the system development life cycle at work.

Disposition

The *disposition* (or *disposal*) phase is initiated to either phase out a portion of a system, or to terminate it completely and end the development life cycle. Once a system is deemed to be obsolete, the first step in this process is to prepare a *disposition plan*. This plan outlines an orderly fashion in which the disposition tasks are to be performed. The plan also addresses who is responsible for each task as well as any time constraints for completing the phase.

One of the most important steps of the disposition phase is data archival. For several reasons, some materials created throughout the life cycle process must be preserved and/or transferred. These materials include data, software components, documentation, and any cryptographic keys associated with the storage of encrypted data.

Life cycle documentation is sometimes referred to as *deliverables*. The main reason for data archival is to facilitate a resurrection of the project if another organization were to bring the system back to life. Depending on the nature of the project, there might also be local, state, or federal policies that need to be observed concerning the preservation of related electronic records.

On the other hand, some materials will need to be destroyed, or *disposed* of. Again, what's saved and what's destroyed depends on the system in question. Software, data, or equipment might be among the casualties of the disposition phase. Hard drives or other media involved in the life cycle might need to be erased. Key words related to the erasure process include degauss, zeroize, sanitize, and overwrite.

It was a good system that had a long life but now it's time to say goodbye. The last step of this, the final phase of the SDLC, is the preparation of a *Post-Termination Review Report*—a sort of eulogy, if you will. This document details all the activities of the disposition phase, what was learned in the process, where the data, software, and equipment may be found, and who to contact to locate the archives. Finally, a short time (usually six months) after the end of this phase, a phase-end evaluation takes place to review the disposition process and make sure everything went according to plan.

SEPARATION OF DUTIES

Separation of duties breaks up the tasks in the development process to ensure that one person does not have too much control over any specific area of development. For example, a programmer will hand off finished code once it is developed. Another team member will then be responsible for implementing it into a production environment. A related process, aptly termed *accountability*, provides a means of connecting individuals with their actions. If there is a collapse of a control in the system, or a security violation occurs, accountability measures ensure that the right person is held responsible.

CERTIFICATION AND ACCREDITATION

This is the bureaucracy of systems development. Here, we have a pair of terms whose subject matter depends on the system being certified and accredited. The terms of the certification/accreditation process are dictated by whichever governing body has set the standards that are applicable. These procedures are carried out during the implementation phase of the SDLC.

Certification

The first half of this duo is mostly concerned with evaluation. Usually carried out by an unbiased outside source, the certification process compares systems with a predefined set of security standards in search of violations. The security features of a system are analyzed to ensure they meet any requirements or laws. It's during this process that the comprehensive documentation created during the SDLC is invaluable. Certification is essentially the first step in accreditation.

Accreditation

The second half of this procedure is where to find the payoff. Also called an *approval to operate*, accreditation gives an organization the permission to put a system into production. Upon accreditation, a system's security features are deemed to be adequate and are formally accepted. Accreditation is usually restricted to a defined period of time, upon which the process must be repeated. Accreditation is also associated with an acceptance by management of any risks involved with operating the system.

PROGRAM LANGUAGES AND EXECUTION

A few terms regarding programming languages and how they are processed need to be clarified. Like English, French or Spanish, each different programming language has its own vocabulary and specific rules regarding grammar (or *syntax*). Let's take a look at the basic categories of languages and how computers read them.

Types of Languages

Here, we'll consider the following types of languages:

- **Machine language:** Also called *machine code* or *instruction code*. This is the form of binary code that a computer can process directly with the help of built-in *instruction sets*. In the beginning, this is how all programs were written. Machine code looks something like this: 1101 0101. Because every kind of computer only understood its own brand of machine code, and writing programs with 1's and 0's produced many errors and took for-*ever*, it became necessary to find another way to program.
- **Assembly language:** Developed to fix the problems of programming in machine code, *assembly language* makes things easier by using symbols that represent binary instructions. For a computer to

process it, assembly code must be converted into machine code, which is done by an *assembler*. Although it simplifies programming a little bit, writing in assembly language is still tedious and prone to typographical errors.

- **High-level languages:** To enable the efficient production of computer programs, many *high-level languages* have been developed that are easier for people to write with. Along with special characters and formulas, this method also makes use of simple English and mathematic terminology, resulting in the most popular and productive way to program. High-level languages decrease the amount of code a programmer must write. Java and C++ are high-level languages. These languages are also *portable*, which means that one does not have to write programs for any particular type of computer architecture. High-level code must also be converted into machine code before a computer can use it, and in the next section we will talk about how this is accomplished.

 Binary machine code is the most raw of any type of programming code and all others must be translated into this format.

Interpretation and Compilation

At the core of a computer, the only language that is understood is machine language. High-level programs must be translated by interpretation or compilation for a computer to execute their commands. These two processes have distinct differences that are explained next.

Interpretation

As you might have guessed, a component called an *interpreter* handles interpretation. Commands are interpreted one line at a time, which is a little slower than other methods. Interpreters can determine the exact line number in the code where an error exists, which is a trick that compiled programs can't do. Java is interpreted. You might recall viewing a Web page that reported a specific line of code that contained an error in its Java script. Interpretation enables this handy method of detecting bugs.

Compilation

Yes, high-level programs are indeed compiled by a *compiler*. A compiler takes high-level *source code* and converts it to *executable code* in one fell swoop. This produces quick-running, OS-specific executable files such as a

Windows *.exe* program. C++ is compiled. Compilers check for errors during translation but because they can't spot every kind of error, debugging can be more difficult. Also, compiled code runs the risk of having malicious code embedded within it. This is hazardous because it can be hard for a programmer to find such dangerous code among a mass of innocent code.

SECURITY CONTROLS

This area of applications development covers the combined concepts, principles, and controls that address the key facets of security throughout the process of systems development and beyond. These controls cover the operating systems, networks, storage methods, applications, and equipment involved in development. Focusing on the security of every aspect of system design, development, and operation, security controls establish common objectives to ensure that the confidentiality, integrity, and availability of the system are upheld to the fullest.

There are numerous tools, standards, and practices that fit into this catch-all category of security-related systems development and application environment controls. We won't expound on every possible scenario that exists in the real world. Therefore, with the objective of Security+ certification in mind, and in unison with the goals of this book, Chapter 7 will conclude with the elements of this group that are most likely to be seen on the exam.

Application environments are divided into *domains*, which are simply groups of objects. *Subjects*, such as users or processes in an application, request access to objects. *Objects* are defined by classifications and sets of rules governing what subjects may do with them, called *permissions*, must be set. In a system, a central component called the *kernel* manages requests for resources. The relationships between object access and subject permissions, and the processes that enable the necessary communications, create security concerns. In the following sections, we'll discuss the tools that ease these concerns, as well as some standards that you should be aware of.

 The kernel is the top-level, authoritative component that controls requests for resources in an application environment.

Process Isolation

Process isolation is where a unique memory address is assigned to each active application. This is done to isolate the processes of individual applications so that they don't modify each other's data. Isolation stops data leakage and also aids a system that is juggling multiple application processes by providing a way to track the whereabouts of each process. Similarly, *hardware segmentation* defines the *amount* of memory an application and its processes can consume. *Resource isolation* separates resources to keep processes from accessing each other's resources.

Least Privilege

This control measure ensures that an application process has only the necessary amount of privileges to accomplish the tasks it is designed to do. This keeps simple processes from having total system access. The least-privilege concept is also applied to users with regard to access control.

Layering

This is a method of dividing application processes into categories (or layers). Tasks are divided based on the role they play in the system. The inner layers perform more sensitive or high-level tasks while the outer layers work with rudimentary tasks. This separation provides a secure, structured way for processes to communicate with each other. In a process known as *data hiding*, layers are made invisible to each other to prevent any intermingling of processes whatsoever. Data hiding is also *called information hiding*.

DATA ABSTRACTION

This is the method of classifying objects in a system. *Abstraction*, which is the underlying basis of OOP, determines a set of admissible permissions and behavioral characteristics for objects.

Reference Monitor

Implemented by the *security kernel*, this is an intermediary software procedure that monitors process requests. When a process requests access to resources (or a *subject* requests access to an *object*), the reference monitor decides if the request should be granted.

Change Controls

The *change control* process exists to provide a management system that handles changes to applications, networks, or any other integral component of a system. The idea is to set up a system of requests and approvals that govern how and when changes are made, not unlike the checks and balances routine outlined by our constitution.

Change controls are comprised of a series of steps that the proposed changes must work their way through before being transferred to a production system. This lesser process of making changes to an established system mirrors the system development process as a whole; changes fall into the maintenance category of the SDLC. With a focus on full documentation to record every step of the process, this practice is implemented as follows:

1. **Request:** By filling out the proper paperwork, a formal request is made.
2. **Analyze:** In this phase, costs are calculated, security-related concerns are taken into account, and an estimated time of completion is gauged.
3. **Review:** Management reviews the proposal and if it looks good, the change is approved.
4. **Develop:** The actual code is added to the application or the existing code is altered.
5. **Test:** Nonproduction testing and quality checks are performed until any existing bugs are worked out.
6. **Change:** The change is implemented into production. If necessary, versions of software will be updated to reflect the change. Change controls protect the availability of a system, which could be disrupted by the implementation of an untested change.

Data/Information Storage

As you know, data and information are stored and archived via different methods, on various types of media. The systems used to retain and record information are categorized under the following headings:

- **Primary:** This is the central memory of a computer. *Primary storage* is referenced directly by the CPU. Primary storage is limited to the amount of memory in a system, making it relatively minimal in size, as well as costly. Chunks of data in memory are given an address that the system calls when referencing them. When a program crashes, the resultant error might mention one of these memory addresses in an indecipherable string of characters on your (now blue) screen. A loss of power can destroy data in memory because it is kept alive by electricity only. This weakness of primary storage is known as *volatility*. *Random Access Memory (RAM)* provides primary storage space.

- **Secondary:** This means that the data stored on hard disk drives or other magnetic media. Secondary storage is non-volatile and it supports primary storage with its larger capacity and lower price tag. Being one of the components in a computer with moving parts, HDDs (hard disk drives) can be prone to failure. Executable data in secondary storage must be moved to primary storage before it can directly interact with a computer's CPU or a device on the system.
- **Real:** When an application has been allotted its own chuck of memory to reside in, it is said to be in *real* storage.
- **Virtual:** Secondary storage devices are used to increase the apparent amount of physical memory in a computer, resulting in virtual memory. To an application, it appears to be the same as primary storage. Through a process called *paging*, data is shifted back and forth between the HDD and primary storage areas. Although virtual memory can greatly increase the capacity for running large applications, the process comes with a reduction in performance.
- **Sequential:** This describes storage that is accessed intermittently, such as a magnetic backup tape.

 Virtual memory is made possible by the secondary storage areas of hard drives. It is used to increase the amount of primary storage.
NOTE

A defined subset of subjects and objects consists of the following:

- Trusted computing base
- Security perimeter
- Reference monitor and security kernel
- Domains
- Resource isolation
- Security policy
- Least privilege
- Layering, data hiding, and abstraction

MODES OF OPERATION

To implement security with regard to what an application process can and can't do, at least two modes of operation are specified. In *supervisor* mode, a process can execute any command it wishes. In *user* mode, it is restricted. This division is made by separate sets of *instructions* for each mode. Each set of instructions has its own *privileges*. This is done so that users or other applications' processes cannot take over a computer's processor.

Applications need total control to do their job, but at the same time, total control must be kept from outside influences. Modes of operation are toggled by the application to provide necessary functions while keeping the system protected. These mode shifts act like a security guard that protects the system's resources. For example, if a user tries to execute a command that requires supervisor privileges, the application switches to supervisor mode and performs the operation. Before giving control back to the user, however, the program will revert the process to user mode. In effect, this distinction of modes protects the memory address space of each application.

Without going any deeper into programming theory, just remember that these modes prevent hostile takeovers by malicious users and they prevent the processes of rogue applications from successfully attacking a system. Supervisor and user modes are defined in a processor's architecture and programs are written to take advantage of this device. The separation of these modes is commonly represented by a set of theoretical *rings*, as seen in Figure 7.4. These rings correspond to the multiple levels of trust defined by the processor's architecture.

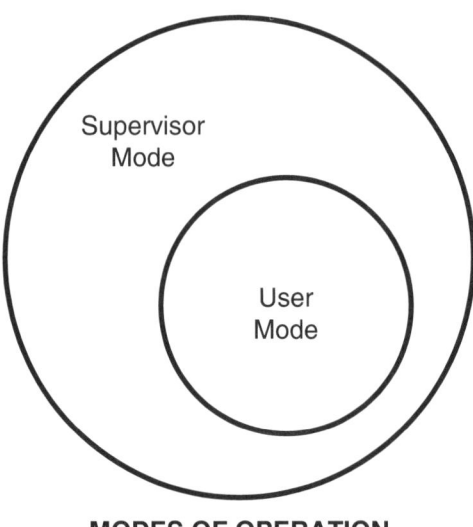

MODES OF OPERATION

FIGURE 7.4   Modes of operation.

TEST TIPS

The following tips have been collected to sum up the topics covered in Chapter 7. They represent the areas of study that deserve the most attention, so memorize away!

- √ A *distributed computing environment (DCE)* is where applications are running across multiple computers in a network. Distributed applications include applets (Java, ActiveX), agents, and objects.
- √ The *nondistributed* environment category of applications presents a threat to the local system. Nondistributed applications include viruses, worms, Trojans, and logic bombs.
- √ *Updates* come in many forms. Know the differences between hotfixes, services packs, patches, and upgrades.
- √ Update = version 1.0 > 1.5. Upgrade = version 1.5 > 2.0.
- √ Staying current is vital. Subscribe to mailing lists to keep up-to-date on security bulletins.
- √ To harden Web servers, use SSL, block unused ports, and disable unnecessary services. For remote administration, replace Telnet with SSH.
- √ To harden e-mail servers, use secured versions of SMTP, POP3, and IMAP. Kerberos and S/MIME can also secure mail traffic. Scan for viruses.
- √ To harden FTP servers, disable anonymous accounts.
- √ To harden DNS servers, use DNSSEC.
- √ To harden DHCP servers, limit DHCP scope, extend lease times, and reserve IPs.
- √ To harden **any** server, apply updates. Use encryption and firewalls. Make use of redundancy techniques such as data backups, secondary servers, RAID arrays, server clusters, mirroring, replication, and backup power.
- √ Memorize the interchangeable database terms. Tuple = row.
- √ A *relation* is a table of rows and columns in a relational database.
- √ *Metadata* is data about data.
- √ *Relationships* aim to link the elements of a database together. Know the difference between one-to-one, one-to-many, and many-to-many

relationships. Also remember that different database models make relationships in different manners.

√ *Views* restrict what is displayed in a database.

√ Both *network* and *hierarchical* databases use *parent-child relationship models*. Children in network databases can have more than one parent. Children in hierarchical databases have only one parent.

√ *Data mining* extracts trends and patterns from a data warehouse.

√ *Aggregation* is a security threat where combined low-level access rights grant access to high-level data.

√ *Inference* is the problem of high-level data being deduced via a portrayal of low-level data. Know the methods that prevent inference attacks.

√ *Polyinstantiation* fights inference attacks by inserting multiple copies of data with differing security levels.

√ *Perturbation* or *noise* prohibits inference attacks by inserting phony data in a database to confuse attackers.

√ *Partitioning* splits up a database to complicate inference.

√ *Covert channels* are exposed links between high-level and low-level data in a multilevel secure (MLS) system.

√ *SQL* is the standard for query languages used by DBMSes. They define the structure of queries.

√ *Knowledge-based systems (KBS)* solve problems by interfacing users with a knowledge base.

√ *Expert systems* are KBSes that are fed rules and actions. They don't learn—they just respond.

√ *Neural networks* are KBSes that simulate biological nervous systems. They are trained and they can learn over time. Also known as *AI (artificial intelligence)*.

√ *SDLC* splits application development into phases. It focuses on documentation, teamwork, and standardization.

√ The *certification/accreditation* process provides an approval to operate.

√ *Machine code* is what all program languages must ultimately be translated into.

√ *Java* is a high-level language that is interpreted or translated one command at a time.

√ *C++* is high-level language that is compiled or translated into an executable file in one process.

√ *Abstraction* classifies objects in OOP.

√ *Reference monitor* monitors subject requests for objects.

√ *Primary storage* is the central memory area of a computer. It is volatile.

√ *Secondary storage* consists of HDDs and the like. It is not volatile.

√ *Virtual storage* is comprised of secondary storage areas. It compliments primary storage.

√ *Sequential storage* refers to intermittently accessed storage areas like backup tapes.

CHAPTER SUMMARY

As you can see, application development relies heavily on standards, teamwork, documentation, and the mechanisms in place to ensure the process follows all set guidelines. While the cooperation of team members is vital during development, management must also make certain that no one person can adversely affect the advancement of a project.

The quality and reliability of a system under development is of utmost importance. Standards that provide separation of duties and other supervisory controls must be used to keep this goal in mind as well as to protect the contributions of each programmer, analyst, and engineer on a team.

Upon reading this chapter and successfully answering all review questions, you should have a good idea of the different computing environments and related applications, application hardening, database structure and security, data warehousing and mining, and the phases of the development life cycle. A basic understanding of application processes, how they communicate, and how they are secured is also essential.

At this point, you are cordially invited to test your knowledge of these topics with the following questions.

REVIEW QUESTIONS

1. **Which of the following types of databases supports the multi-parent concept?**

   ○ A. Network.
   ○ B. Hierarchical.
   ○ C. Relational.
   ○ D. Flat.
   ○ E. Data mining.

   Correct answer = A

   Children in network databases can have multiple parents. The hierarchical model has children with only one parent. Relational databases don't use the parent-child approach to making relationships. Flat databases also do not use the parent-child approach to making relationships. Data mining is a process that extracts trends and patterns from a data warehouse.

2. **Which of the following best describes cardinality?**

   ○ A. The number of files contained in a relation.
   ○ B. The number of fields contained in a column.
   ○ C. The number of tables contained in a row.
   ○ D. The number of rows contained in a relation.
   ○ E. The number of attributes contained in a field.

   Correct answer = D

   Cardinality is the number of rows (or tuples) in a relation. Answers A, B, C, and E are invalid.

3. **Which type of agents can roam networks, make copies of themselves, and gather information?**

   ○ A. X agents.
   ○ B. Paleolithic agents.
   ○ C. Polymeric agents.
   ○ D. Mutated agents.
   ○ E. Mobile agents.

   Correct answer = E

Opposed to static agents, mobile agents can copy themselves, roam networks, and gather information. These characteristics enable mobile agents to do things such as deliver viruses. Answers A, B, C, and D are invalid.

4. Which type of urgent update is released to repair bugs quickly?

○ A. Upgrade.

○ B. Hot-fix.

○ C. Stable fix.

○ D. Service pack.

○ E. Rapid fix.

Correct answer = B

Hot-fixes are released when a quick remedy is needed for a bug. Upgrades indicate a shift in product versions as in 4.2 to 5.0. Service packs are collections of fully tested hot fixes and other utilities/updates. Answers C and E are invalid.

5. Which type of storage is used to create virtual memory?

○ A. Secondary.

○ B. Primary.

○ C. Long-term.

○ D. Sequential.

○ E. Real.

Correct answer = A

Secondary storage devices such as HDDs are used to create virtual memory, which enables a system to run large applications that its primary storage area doesn't have the capacity for. Primary storage is the physical memory of a computer. Sequential storage refers to tapes or other storage devices that are periodically referenced. Real storage refers to the resident areas of memory that applications are assigned. Answer C is invalid.

6. Which of the following is an incorrect matching of database terms?

○ A. Relation = table.

○ B. Attribute = column.

○ C. Tuple = row.

○ D. Field = attribute.

○ E. Record = attribute.

Correct answer = E

From traditional database terminology, records are synonymous with rows and tuples, not attributes. Refer to Table 7.1 to see that answers A, B, C, and D represent proper match-ups of interchangeable database terminology.

7. **Which of the following methods is *not* used to fight inference attacks?**

○ A. Abstraction.

○ B. Perturbation.

○ C. Noise.

○ D. Polyinstantiation.

○ E. Partitioning.

Correct answer = A

Abstraction is the process of classifying objects in object-oriented programming. Perturbation and noise are synonymous terms for the process of inserting phony data into a database to fight inference attacks. Polyinstantiation combats inference attacks by inserting multiple instances of data into a database with different security levels. Partitioning protects against inference attacks by splitting a database into multiple parts.

8. **Denormalization can have an adverse effect on which aspect of data?**

○ A. Data regularity.

○ B. Data confidentiality.

○ C. Data integrity.

○ D. The time it takes to access data.

○ E. Data magnification.

Correct answer = C

Denormalization can affect data integrity because it introduces the duplication of data into a database. Denormalization is not related to data confidentiality and its purpose is actually to reduce the time it takes to access data, which is a positive effect. Answers A and E are invalid.

9. **Which of the following is a tool used in Customer Relationship Management (CRM) techniques to find patterns in online behavior?**

   ○ A. URL structuring.

   ○ B. Web mining.

   ○ C. OLTP mining.

   ○ D. ODBMS.

   ○ E. CORBA.

   Correct answer = B

   Web mining employs content, usage, and structure mining to uncover trends in online behavior. ODBMS is an object-oriented database management system. CORBA is an object request broker. Answers A and C are invalid.

10. **In an MLS-based system, which of the following describes a threat exposed by linking high-classification data to low-classification data?**

    ○ A. Security label.

    ○ B. Covert channel.

    ○ C. Security policy.

    ○ D. Mandatory access controls.

    ○ E. Metadata listing service.

    Correct answer = B

    A covert channel is an exposed link between high-classification data and low-classification data. A security label specifies the security classification of an object. A security policy defines security labels. Mandatory access control is an authentication method. Answer E is invalid.

11. **Which process translates programming code one command at a time?**

    ○ A. Relation.

    ○ B. Interpretation.

    ○ C. Decryption.

    ○ D. Assimilation.

    ○ E. Compilation.

    Correct answer = B

Interpretation translates programming code one command at a time as in Java interpretation. A relation is a table of rows and columns in a relational database Decryption is the cryptographic process of decoding cipher text. Compilation translates programming code into a compiled executable all at once. Answer D is invalid.

12. **Why is accreditation a necessary step before certification?**
    - ○ A. To make sure all security features are operating properly.
    - ○ B. Because any risks associated with operating a system must be accepted by management before the certification process can begin.
    - ○ C. Because it's a law.
    - ○ D. Because you need an *approval to operate* before the certification process can begin.
    - ○ E. It isn't. Certification comes before accreditation.

    Correct answer = E

    Certification actually comes before accreditation. The end goal of the certification/ accreditation process is to attain an *approval to operate*. Answers A, B, C, and D are invalid.

13. **Which type of programming language must all other languages ultimately be translated into?**
    - ○ A. Slag code.
    - ○ B. Assembly code.
    - ○ C. Machine code.
    - ○ D. Java code.
    - ○ E. Object code.

    Correct answer = C

    Machine code is the "native tongue" of computers and all other programming languages must eventually be translated into binary machine code. Logic bombs, a type of malicious program, contain slag code. Assembly code is programming language that uses symbols instead of raw binary data and it must be assembled into machine code. Programming in the high-level language of Java produces Java code. Object code is an OOP term.

14. **Which of the following defines the process of accessing high-level information by merging collections of low-level data?**
    - ○ A. Aggregation.
    - ○ B. Polyinstantiation.
    - ○ C. Inference.
    - ○ D. Abstraction.
    - ○ E. Assemblage.

Correct answer = A

Aggregation is a condition where a combined assortment of low-level access rights provides access to information of a higher sensitivity. Polyinstantiation combats inference attacks by inserting multiple instances of data into a database with different security levels. Abstraction is the process of classifying objects in object-oriented programming. Answer E is invalid.

15. **In a database, attributes are represented as _____.**
    - ○ A. Vertical named rows.
    - ○ B. Horizontal named columns.
    - ○ C. Vertical named columns.
    - ○ D. Horizontal named rows.
    - ○ E. None of the above.

Correct answer = C

Attributes, which describe an object or entity in a database, are represented as vertical named columns. Tuples are represented as horizontal named rows. Answers A, B, and E are invalid.

16. **During the SDLC, which phase needs to be documented fully?**
    - ○ A. Feasibility.
    - ○ B. Implementation.
    - ○ C. Maintenance.
    - ○ D. Disposal.
    - ○ E. All of the above.

Correct answer = E

During the SDLC, every step of the way should be fully documented. This includes all phases. Therefore, answers A, B, C, and D are all correct.

17. **Which database security mechanism prevents a database from displaying certain data to specific users or groups?**

○ A. Filters.

○ B. Views.

○ C. Queries.

○ D. Degrees.

○ E. Metadata.

Correct answer = B

Views are used to control who can change or view individual elements of a database—they can be set for individual users or entire groups. Filters are criteria used to include (or exclude) specific information in the results of a database query. Queries are commands used to fetch and display requested information from a database. A degree specifies the number of attributes in a relation. Metadata is data about a database.

REFERENCES

http://www.research.ibm.com/iagents/ is IBM's Intelligent Agents Project Web site. It contains facts about mobile Internet agents and rule-based programming techniques. Be sure to check out their main research pages as well.

http://searchdatabase.techtarget.com leads to all things database—White Papers, tech advice, database-related news, and more. Among other topics, TechTarget maintains a database-specific search engine and central information repository.

http://java.sun.com/sfaq is offered by Sun Microsystems, Inc. It is a resource for security-related Java issues and has an informative FAQ section on applet security.

CHAPTER 8

LAW, CRIMES, ETHICS, AND INVESTIGATION

IN THIS CHAPTER

- Law
- Computer Crime
- Ethics
- Investigation
- Test Tips
- Chapter Summary
- Review Questions
- References

A computer or network security-related book would not be complete without a chapter that specifically focuses on computer-related law, ethics, fraud, computer crimes, and investigation. Although it is probable that you will only face general questions regarding law, ethics, investigation, and computer crimes, it is important for you to understand these topics in order to have a well-rounded foundation as a security specialist. It is also important that you have the ability to answer even the general questions on the Security+ exam. In Chapter 6, you were introduced to some of the topics that are also included in this chapter to better prepare you for the subject matter contained in it. Remember, the most important information pertaining to the exam will be identified for you.

LAW

As you are probably aware, similar to the topic of security, the topic of law and legal affairs covers a very broad range of subjects and information. In order to understand the laws that are relevant to computer-related crimes and prepare you for the subject matter later in this section, it is important for you to understand the legal terms that are described next.

 Several current security-related certification exams refer to these terms as *intellectual properties*. Just for the record, intellectual properties can include subjects such as software piracy, trademarks, copyrights, patents and patent infringements, and trade secrets. However, before we discuss the intellectual properties, it is important for you to have a general understanding of the major types of laws that exist. They are listed as follows:

- **Criminal law:** This type of law protects society from individuals or groups that violate laws enacted by the government. With this type of law, the government prosecutes those who have committed what it deems as crimes. With criminal law, the government or state will appoint a prosecutor to file a suit against such crimes as murder or rape. These crimes are deemed as felonies or misdemeanors.

- **Civil law:** This type of law has to do with wrongful doings between individuals or between businesses and individuals. It typically results in some sort of loss or damage. A civil case is brought forward when an individual or company wishes to resolve the loss or wrongful doing that has occurred due to the negligence or misconduct of another. With civil law, compensation for loss is typically requested.

- **Administrative law:** This type of law pertains to government agencies, organizations, and offices. It regulates how the agencies should conduct their affairs and business dealings.

Next, we will discuss trademarks, patents, copyrights, and trade secrets.

 It is likely that the Security+ exam will expect you to know the difference between trademarks, patents, copyrights, and trade secrets.

TRADEMARK

A *trademark* is a word, phrase, title, logo, symbol, design, letter, number, or other object that it used to uniquely identify a company, product or service. Just about anything you can buy today has a trademark. Computers, software, network appliances, and even ISPs use registered trademarks. A *registered* trademark is trademark that has been granted for a particular product or service by the federal government. If another person or company creates a product or service and places a label or title on that product or service that is very similar to a registered trademark, the trademark's registered owner can possibly sue that individual or company.

PATENT

A *patent* is a privilege or right of use that is specifically assigned by government to the creator, developer, or owner of a process, mechanism, service, or product. When something is patented, it is protected by law from being copied or misused by anyone other than the patent holder. In simple terms, if you develop a product and receive a patent from the U.S. Government Patent Office regarding that product, you have exclusive use of manufacturing and reproducing that particular product.

COPYRIGHT

A *copyright* is the right to create and sell what is exclusive to the creator or owner of the copyright. Copyrights are usually associated with such creations as books, articles, movies, and music. For example, an author can be the creator or inventor of an idea or expression. The author or publisher of the idea or expression will typically have the material registered and copyrighted in order to protect the idea or expression from being modified, copied or sold by someone else. In order to obtain a copyright, one must register for the copyright with the federal copyright registry.

TRADE SECRET

A *trade secret* is confidential company or business information or devices such as a secret formula, code, database, device, or other product whose effectiveness and profitability are based on its secrecy. In other words, in order for a company to have or maintain an advantage over a competitor, its trade secrets or company secrets must not be divulged.

 For the exam, know that a trade secret is proprietary company information whose secrecy is essential to the health and profitability of a company.

COMPUTER LAWS

As mentioned in the opening paragraph of this chapter, there are specific laws that have been enacted which relate to the protection of computer security and privacy. The demand for more secure electronic business transactions and the need for increased protection regarding personal privacy are on the rise. More computer-related security laws and stricter punishment, for those who break them, are needed to protect the privacy and welfare of businesses and individuals in this highly technical age.

There are many new computer privacy laws currently being considered by government. Next, you will be provided with a list of valid security and privacy laws. It is not likely that the Security+ exam will expect you to know them all. However, you never know what CompTIA might have in store as the exam popularity grows and different questions are added to the exam. Look closely at the following laws, you will notice they are relative and apply to many of the topics discussed in this book:

- **1974 Federal Privacy Act:** This act protects the public by insuring that personal information held by federal agencies is not accessible or released without consent.

- **1984 First Computer Security Act:** This act states that unauthorized access to or destruction of federal computing systems or information is a felony.

- **1986 (amended 1996) Computer Fraud and Abuse Act:** This act better defined the first computer security law created in 1984. It was updated in 1996 to include the identification of three new computer crimes: use of a federal interest computer to further intended fraud; the altering or destruction of information on a federal interest computer that causes $1,000 in loss or results in medical treatment; and the trafficking of computer-related passwords that allows access to government computers.

- **1986 Electronic Communications Privacy Act:** This act prohibits eavesdropping by way of wire or oral communications. It addresses the important issue of information being illegally obtained without consent. In simple terms, in order for someone to monitor your communication or information, they must first be granted explicit permission from the legal or court system. For example, police or federal agencies must obtain legal permission to wire tap or carry out surveillance on a suspected criminal.

- **1987 Computer Security Act:** This act requires federal agencies to identify critical information systems, provide computer security training to employees, and develop documented plans to secure all systems.

- **1994 U.S. Communications Assistance for Law Enforcement Act:** This act calls for all communications carriers and companies to provide the ability for wire tapping to be used.
- **1994 Computer Abuse Amendments Act:** This act is better defined by the phrase "federal interest computer" to include a system or computer that is used for interstate commerce or communications. This act is also updated to include worm and virus laws.
- **1996 U.S. Economic and Protection of Proprietary Information Act:** This act covers computer security laws and punishment relating to corporate enterprise as well as industrial espionage.
- **1999 Gramm-Leach-Bliley Law:** This law was enacted for the further protection of nonpublic personal information. Its main focus is to ensure that financial institutions have an obligation to protect the privacy of their customers by implementing and supporting technical, administrative, and physical safeguards.

If you are interested in learning about the newest computer-related laws including copyright, trademark, and other legislation, you should definitely visit *http://www.complaw.com*. Additional legal information and resources can be found at *http://www.findlaw.com* and *http://www.usdoj.gov/criminal/ cybercrime/cclaws.html*.

COMPUTER CRIME

Computer systems and software have helped the world become a highly productive place. We can educate ourselves with just about any topic in minutes using our little windows to the world. We can reach thousands of people with just a couple of clicks. We can make money, get a job, and do other productive things such as quickly help other people in under privi-leged countries use money and computers.

Computers and the Internet have provided a gift to us human beings. Or is the flip side of the coin true? Are computers and the Internet the battle-ground on which future wars will be waged?

It is becoming apparent, in this technical age in which we live, that com-puters are becoming the weapons of choice for criminals, terrorists, dis-gruntled geniuses, and highly educated little kids that are bored with electronic games.

Based on the recent dramatic increase in computer-related cyber crimes, it is likely that the future CompTIA Security+ certification exam might include an entire section of questions dedicated to computer crime and associated computer crime attacks. For the current Security+ exam, it is important for you to have a basic understanding of the following attack categories and computer crimes, which are considered illegal.

FUN ATTACKS

Computer *Fun attacks* are computer-related attacks or breaches of computer security generally committed by younger people or *script kiddies*—college students or the average Joe with a little more knowledge than they should have, sitting at home experimenting with a computer system and a hacker's guide usually purchased at a local bookstore. Fun attacks are not typically meant to cause damage or harm. They are not generally used as a means to steal or gain specific information that would jeopardize the welfare or health of a business. Instead, these types of attacks are generally all about ego.

The person carrying out the Fun attack is usually interested in seeing if they can first break into a system or network, then seeing how far they can get. The goal of the Fun attacker is of a boastful nature. "What can I get?" "How far can I get?" and "Who can I tell how special I am for getting there?" are usually the interests of this individual or group. If you did your homework in the earlier section regarding computer law, you are aware that unauthorized entry into a personal or business system or network is a crime. Therefore, those waging Fun attacks that mean no harm are still breaking the law and committing punishable crimes. In simple terms, if you go into a locked store and do not steal anything, you are still breaking the law whether or not you meant any harm to anyone.

GRUDGE ATTACKS

A *grudge* attack is an attack on a program, system, or network that is usually initiated by a disgruntled employee or former employee who feels they have a score to settle with a certain company or employee. These types of attacks are usually launched from remote locations using existing VPN connections, or in the form of a particular type of malicious code, such as a logic bomb. It is difficult to protect inside information and systems from newly released or currently disgruntled employees. Strict company policies and fast communication methods should be in place to quickly notify network and security personal of employees who have left the company or employees who exhibit particular behavior.

BUSINESS AND FINANCIAL ATTACKS

It makes sense that hackers and would-be thugs often target profitable businesses and financial institutions for information and possible profit. It is highly important for these types of businesses to employ strict policies and use only the most effective security defense mechanisms. Because these types of organizations are high profile and are generally accessible by the public for commercial purposes, they are tempting targets for cyber criminals.

TERRORIST ATTACKS

Recent terrorist actions have prompted governments and individuals to scrutinize their physical as well as logical defense mechanisms. It is obvious that computerized terrorist attacks will eventually target information systems and Enterprise networks more frequently. Terrorists use computers and software to manipulate funds, trade information, and carry out other tasks that inevitably result in destruction and crime. Important legislation is enacted and more legislation is pending to protect the welfare and privacy of certain governments. An important act that addresses many of the growing concerns regarding the protection of information systems by deterring and obstructing terrorism is the U.S. Patriot Act. Public information is available regarding the Act at *http://www.cdt.org/security/010911response.shtml.*

DATA DIDDLING

Data diddling is one of the most commonly committed computer crimes of the day. An internal company employee typically carries out this type of crime. It involves the changing of data or information before or as it is entered into a system or program. The data is usually changed back to original form after it has been processed or calculated. Data diddling is common in financial institutions, businesses, schools, and government. It costs companies, organizations, and ultimately taxpayers a fortune every year to cover the losses of this difficult-to-detect attack.

 Data diddling is by far the most common form of computer-related crime. Employees of companies and organizations commit more computer-related crimes than any other group.

SALAMI ATTACKS

A *salami* attack is a computer-related attack with intention of making a financial gain using very small increments of information and money that usually go unnoticed. For example, if a program is written to deduct 10 cents from the separate automatic payroll deductions of 400,000 employee paychecks before the paychecks are deposited into the proper bank accounts, the program writer can devise a scheme where the selected amount is deposited into a separate account and make quite a profit. Most of the employees will not notice that they were missing 10 cents. It would be possible for utility companies to carry out a similar salami attack. For example, say 700,000 people were "accidentally" overcharged $1.00 on their utility bill for one month. Most people would not notice a thing. The amount of $700,000 would be a big mistake, wouldn't it?

SOFTWARE PIRACY

Software piracy can be defined as the illegal duplication, use, and distribution of software. It is now estimated that 40 to 50 percent of all currently used software is pirated or stolen. Chances are that you, the company you work for, or the school you are enrolled at are using illegal software in one form or another.

Many private companies that sell computers or *clones* will build computers and load one licensed copy of an operating system on many systems. They make an illegal profit by only purchasing one legal license. When you purchase a system from a seller or computer dealer, make sure you get the software and a legal license from the manufacturer to use that software.

The best way to fight piracy is to use only licensed software and report uses of counterfeit or illegal software usage to the original manufacturer of the software.

There are strict punishments for those convicted of software piracy. Stealing software is no different than stealing anything else. In other words, when you burn a copy of that really neat game, top-ten music CD-ROM, or copyrighted schoolbook, you are committing a felony punishable by up to five years in prison.

ESPIONAGE

Espionage is considered the act of spying on someone or something with the intent of gaining secret, personal, or classified information. More specifically, computer and information related espionage is the act of spying on computer systems, networks, and stored information with the intent of obtaining confidential information. Where it applies, most governments and financial business institutions are required by law to institute certain intrusion detection systems, monitoring and other devices in order to protect themselves from unauthorized access as well as espionage.

EMBEZZLEMENT

Embezzlement is the illegal use of or stealing property that belongs to someone else that has been entrusted to your care. For example, someone trusts you to take care of his or her bank accounts. You, in turn buy a new a car with his or her money. Embezzlement crimes typically involve an element of trust of confidence. Many embezzlement cases involve lawyers, bank employees, business leaders, trustees, and agents. These are just a few examples of those entrusted with Due Care and moral responsibilities.

There are many laws in place to protect society from embezzlement. Company policies often include statements that potential employees must read and sign regarding embezzlement and the punishments that exist for those caught embezzling.

FRAUD

Fraud is the intentional misrepresentation of the truth in order to gain a business edge, financial profit, or something considered valuable. Fraud can also be defined as trickery, deception, and lying in order to position oneself for illegal gain.

The Internet provides a huge playing field for scammers and cheaters to carry out fraudulent activities. Fake auctions, deceiving providers of goods and services, false advertisements, and many other illegal fraudulent booby traps await innocent web surfers who believe what they read and what they click on.

Fraudulent activities at work should be reported to Human Resources and company management. Fraudulent activity on the Internet should be reported to your Better Business Bureau and possible local or state government or representative.

The following public Internet site contains some excellent information that can help you become better educated and assist you with avoiding computer fraud and other threats: *http://www.techtv.com/cybercrime/aboutus/story/0,23008,3339221,00.html.*

ILLEGAL INTERNET CONTENT

Certain types of information posted on the Internet is considered offensive and in many cases, illegal. It is important that illegal content found on the Internet first be reported to your Internet service provider (ISP), then to local, state, and federal government agencies. The following represent some illegal as well as potentially illegal situations and information that should be reported:

- Cyber harassment or stalking
- Illegal activities, such as bomb threats or terrorism
- Threatening hate crimes
- Computer hacking
- Physical threats
- Child pornography
- Illegal advertisements
- Internet scams

CYBER STALKING

Cyber stalking is defined as the continuous computerized harassment of another person. A cyber stalker will typically stalk or harass others over the Internet using e-mail, Instant Messaging, or other forms of electronic communication. Many ISPs and Instant Messaging software programs offer the ability to block and report potentially threatening users that seem harmful or display threatening behavior.

Cyber stalking is not a legal definition and is not technically considered a crime. However, if a potentially threatening person makes personal threats and uses other threatening means such as phone calls, physical threats, or physically stalks another, electronic evidence trails and documentation might be considered admissible if deemed relevant.

To be on the safe side, any potentially threatening event or person should be reported to your ISP and local authorities.

CYBER PORNOGRAPHY

The Internet was originally created as a communications tool for government use. It then grew into an awesome and vast educational tool. Over time, it has become one of the largest distribution points for free as well as commercialized pornographic material. Congress is currently working on developing and passing laws to better protect privacy and unwanted advertisements that display pornographic material. If you are concerned with children under the age of 18 years of age viewing offensive material and information on the Internet, you should invest in software that blocks access to this type of material. Most ISPs offer the ability to shield or block your system and browser from accessing or displaying unwanted pornography.

Computer crimes are separated into two categories: crimes that are carried out against a computer and crimes committed using a computer.

ETHICS

Ethics are defined as a set of principles or rules that govern one's morals and actions as they apply to one's duty. There are ethics that apply to just about every topic under the sun. There are ethics that apply to certified security specialists, doctors, lawyers, publishers, teachers, use of the Internet, use of business systems, finances, and many other occupations and subjects. Our focus is targeted at ethics that pertain to computer usage, the Internet, business practices, and security certification.

COMPUTER ETHICS

Most "morally sound" people have a pretty good idea of the basic ethics that should be considered and implemented when using a computer and the Internet at home, or when following computer usage policies and practices in the workplace. If you become a security certified specialist it will be important that you are educated with the "Ten Commandments of Computer Ethics" as created by The Computer Ethics Institute and ethics practices stated by the Internet Architecture Board (IAB). It is also important that you are familiar with RFC (Request For Comments) 1087 that pertains to ethics and proper use of the Internet.

It is important for people to consider the rights of others when using computer systems and the Internet. Privacy, laws, and basic common sense should be used as tools to help guide those who need a little help with computer-related ethics. The Computer Ethics Institute does a great job explaining some of these ethics that should be considered.

The "Ten Commandments of Computer Ethics" according to the Computer Ethics Institute are as follows:

1. Thou shalt not use a computer to harm other people.
2. Thou shalt not interfere with other people's computer work.
3. Thou shalt not snoop around in other people's computer files.
4. Thou shalt not use a computer to steal.
5. Thou shalt not use a computer to bear false witness.

6. Thou shalt not copy or use proprietary software for which you have not paid.

7. Thou shalt not use other people's computer resources without authorization or proper compensation.

8. Thou shalt not appropriate other people's intellectual output.

9. Thou shalt think about the social consequences of the program you are writing or the system you are designing.

10. Thou shalt always use a computer in ways that insure consideration and respect for your fellow humans.

© 1991. Computer Ethics Institute

For more information regarding the computer ethics and the Computer Ethics Institute, visit the Computer Ethics Institute at *http://www.brook.edu/its/cei/cei_hp.htm.*

INTERNET ARCHITECTURE BOARD (IAB)

The *Internet Architecture Board (IAB)* is the governor or advisor of the Internet Society. Its primary concern is to provide advice concerning Internet evolution. The three main purposes that the IAB serves are as follows:

- Overseeing the activities of the Internet Engineering Task Force (IETF).
- Overseeing the implementation of Internet standards and their processes.
- Overseeing, managing, and publishing Request For Comments (RFCs).

What does all of this have to do with ethics? The answer: everything. The IAB statement of policy describes the Internet, explains why it exists, and tells how it should be used. The IAB policy specifically states, "Internet Activity should be treated as a privilege." The following information is provided to the public by the IAB. The quotes that you will see next are directly form the IAB Statement of Policy and RFC 1087 that pertain to ethics and the Internet. (Source of these quotes is *http://www.cis.ohio-state.edu/cgi-bin/rfc/rfc1087.html.*) You should already be aware that an RFC is a document that provides information or an approved Internet Engineering Task Force (IETF) Internet standard.

The Internet is a national facility whose utility is largely a consequence of its wide availability and accessibility.

Irresponsible use of this critical resource poses an enormous threat to its continued availability to the technical community.

The U.S. Government sponsors of this system have a fiduciary responsibility to the public to allocate government resources wisely and effectively. Justification for the support of this system suffers when highly disruptive abuses occur. Access to and use of the Internet is a privilege and should be treated as such by all users of this system.

The IAB strongly endorses the view of the Division Advisory Panel of the National Science Foundation Division of Network, Communications Research and Infrastructure which, in paraphrase, characterized as unethical and unacceptable any activity which purposely:

- *Seeks to gain unauthorized access to the resources of the Internet*
- *Disrupts the intended use of the Internet*
- *Wastes resources (people, capacity, computer) through such actions*
- *Destroys the integrity of computer-based information*
- *Compromises the privacy of users*

If you are interested in becoming further educated with RFC 1087 and Internet ethics, please visit *http://www.cis.ohio-state.edu/cgi-bin/rfc/rfc1087.html.*

INVESTIGATION

In order for any computer-related crime evidence to have any chance of holding up in a court of law, certain conditions, rules, and other criteria must be in place and followed. In most cases, professional advice as well as professional evidence collection methods are required for evidence to be considered admissible, relevant, and substantial.

In Chapter 6, you learned the importance of computer forensics, evidence collection, and evidence preservation. It is important that you keep your knowledge of that subject matter in mind as we now focus on developing an overall understanding of investigation preparation and rules to follow when conducting an investigation.

CONDUCTING AN INVESTIGATION

Conducting an investigation within your company requires that certain criteria already be established. First and foremost:

- A company investigation committee should be established.
- An enterprise or corporate investigation team should include all of the appropriate staff. This can include network security, building security, management, Human Resources, accounting departments, and other appropriate groups.
- The committee should anticipate the likelihood of possible computer crime and prepare for such events by preparing for and establishing the following:
  - Creating a liaison with law enforcement an other emergency response agencies.
  - Creating a procedure that documents how and when law enforcement agencies will be notified.
  - Establishing documents and procedures that specify how the various computer crimes will be handled and reported. For example, will the FBI need to be notified? Will a professional forensics team need to be consulted?
  - Establishing documentation that specifies how the specific investigation will be handled and carried out.
- The committee must ensure that evidence is handled properly.
- The committee must be prepared for retaliation as a result of an investigation.
- The committee must be familiar with U.S. federal requirements as they pertain to reporting and handling investigations.

It should always be anticipated that computer crimes will happen. The question is when will they happen? You should prepare for computer-related crimes as you would prepare and protect your networked environment from computer virus attacks. You should have protection and an organized plan way ahead of time.

The frequency of computer crime is expected to increase as our society becomes more and more dependent on computer technology. In order to best protect and react to crimes committed within your organization, it is critical that your company employees are prepared and "investigation smart."

CATEGORIES OF EVIDENCE

It is most important that evidence obtained during an investigation is relevant and admissible in court. Although it is unlikely that the following descriptions of the types of evidence will not appear on the exam, it is important for you to know what they are if you are considering a career in security or investigation:

- **Physical:** This is evidence that you can touch. It is real. In other words, it exists physically. For example, computer forensic evidence is considered physical evidence. Fingerprints are considered physical evidence.
- **Testimonial:** This is evidence obtained through testimonials of witnesses or possible confessions by suspects. Typically, this type of evidence is obtained from those who have information regarding who committed a crime, how they committed a crime, and where they might be located.
- **Demonstrative:** This type of evidence can include photographs or drawings of a crime scene or possible crime scene. Typically, these photographs and drawings are used to recreate or display where and how a crime occurred.

CHAIN OF EVIDENCE CUSTODY

The *chain of evidence custody* is a documented report that identifies who has custody of evidence from beginning to end. It is important to know who has custody of the evidence at hand to ensure accountability. If the chain of evidence custody is broken, or if the evidence is ever misplaced, it is likely that the evidence will prove useless to an investigation and will most likely not be admissible in a court of law. Knowing who has custody of the evidence at all times is critical to the investigation process.

The chain of evidence custody starts when the first piece of evidence is obtained. The chain of evidence custody continues until the evidence is no longer needed. The holder or owner of the evidence in the chain of evidence custody is called an *evidence custodian*. Each person that handles his or her part of the evidence is responsible for the welfare of the evidence. They are accountable and responsible for evidence preservation. Evidence collection and preservation was detailed in Chapter 6. By now, it should be quite apparent to you that the proper chain of evidence custody and the proper evidence collection and preservation techniques and guidelines are going to be a big part of the Security+ exam.

ENTICEMENT

The term *enticement* is defined as the practice or act of alluring, tempting, or attracting someone or something into doing something. For our Security+ focus and study, enticement is related to luring an intruder or hacker into leaving a trail or evidence behind that can be used to prove that they have broken the law. A honey pot can be used as a perfect example to demonstrate this point.

In Chapter 4, you learned that honey pots are used to attract hackers and crackers. A honey pot is basically an unprotected system with no applied patches, operating system updates, or firmware updates that is used to attract, trap, and identify possible attackers. The honey pot that is a system that is monitored, audited and logged. It is used to entice or lure the hacker. A better name for a honey pot might be *mousetrap*.

ENTRAPMENT

Entrapment involves luring or pursuing someone into committing an illegal act that they, the individual, had no original intention of committing. More times than not, entrapment seems to be associated with federal, state, or local officials such as police offices forcing civilians to commit an act or crime that they had not intended to commit.

Concerning computer-related security issues, entrapment has become more of a defense plea or mechanism that cyber criminals are using use as a result of being caught with their hands in the honey pot jar. In other words, a hacker who has been identified and reported after attacking a honey pot server might attempt to argue a case of entrapment as a legal defense.

HEARSAY

Hearsay is evidence that is not generally admissible based on the fact that it is gathered from second hand sources and not directly tied to a first hand specific witness. Hearsay is not considered fact and usually holds little value. However, there are some exceptions where hearsay evidence might be justified and possibly used as admissible evidence. Here are a few possible exceptions:

- If evidence is collected during the regularity of consistent business routines and direct witnesses are involved.
- If evidence is presented by a person with knowledge of records.
- If evidence is presented by a person with knowledge.

- If evidence is gathered close to the time of the actual criminal act.
- If evidence is in the possession of a witness on a regular or routine basis.

SURVEILLANCE

In order to monitor one's activities physically or logically you must let them know you are doing so or have a warrant that allows you to record their activities. Most businesses monitor entryways, exits, and access to secured locations through the use of Closed-Circuit Television (CCTV). Most Network administrators monitor files, folders, and other shared resource access activity through the use of auditing and logging. It is important to keep in mind that most companies include a policy that states that employees will be monitored for their own protection and for the protection of the company's resources.

Law enforcement agencies are required to obtain warrants and other permissions before monitoring and recording suspicious activity. Laws regarding legal surveillance and monitoring must be followed or the evidence obtained for an investigation may not be admissible in court.

MOTIVE, OPPORTUNITY, AND MEANS (MOM)

Investigating a criminal offense, whether it is computer-related or not, often includes putting yourself in the shoes of the suspected criminal in order to figure out why and how they committed a specific crime. Information security specialists and several security examinations (including the Security+ exam) use the acronym *MOM* to describe the "why, when, and how" of computer crime crimes. Don't be surprised if it happens to pop up in a question. Here's what you have to know concerning MOM.

Motive

The question here is, "Why did someone commit a crime." People commit computer-related crimes in order to obtain money, steal information that can be sold, or obtain trade secrets that can give a particular business an advantage over the competition. Some simply inflate their egos with a successful hack of another's system. As long as valuable information or services exist, it is likely that someone will be motivated to obtain it illegally.

Opportunity

The *opportunity* to commit a computer-related crime and be successful at it has increased over time. As the number of systems and people connected to the Internet grow, so does the opportunity to infiltrate weak or unprotected systems. Many systems, both at home and in the workplace, are vulnerable mainly due to lack of resources and security education. Since the dot.com fallout, many businesses have been forced to cut back on the staffing, training, and purchasing of equipment and software that is essential to the protection of an internal network. This can leave the door open for the many who wish take advantage of unprotected networks and systems.

Means

Committing a computer-related or Internet crime is quite easy these days. One can download free scripts from the Internet that can be used quite easily as hacking tools or make a quick stop at the local book store and pick up a "how to hack" instruction book. The tools used to break into systems illegally are readily available and as younger computer educated generations evolve, more computer-related crimes will occur.

COAST (COMPUTER OPERATIONS, AUDIT, AND SECURITY TECHNOLOGY)

COAST is a computer investigation laboratory project that exists at Purdue University. It is one of the largest security research groups in the world. They provide a wealth of valuable information regarding computer investigations on their Web site. In fact, you can even search computer-related crimes and investigations by country or state. It is highly advisable that you visit their Web site at *http://www.cerias.purdue.edu/coast/*.

Don't be surprised if you see reference to COAST while taking the exam.

TEST TIPS

As a reminder, the Test Tips sections included at the end of each chapter are extremely important and geared toward helping you focus on the important points within each chapter. Please make sure you understand the following tips and apply your knowledge gained from these tips and the rest of the chapter when answering the review questions at the very end of the chapter.

√ Employees of companies and organizations commit more computer-related crimes than any other group.

√ The *Gramm-Leach-Bliley law* focuses on ensuring that financial institutions have an obligation to protect the privacy of their customers by implementing and supporting technical, administrative, and physical safeguards.

√ A *trade secret* is proprietary company information whose secrecy is essential to the health and profitability of a company.

√ Computer *fun attacks* are computer-related attacks or breaches of computer security generally committed by younger people or *script kiddies*.

√ Terrorists use computers and software to manipulate funds, trade information, and carry out other tasks that inevitably result in destruction and crime.

√ The *U.S. Patriot Act* focuses on preventing, deterring, and obstructing terrorist attacks.

√ A *salami attack* attempts to achieve financial gain by stealing small amounts of information or money that usually goes unnoticed.

√ *Hackers* can be enticed or lured to unprotected systems called honey pots. Once the intruder has accessed the system, Intrusion Detection Systems, logging, and other tools can be used as an attempt to identify the intruder and use any trails left behind as evidence.

√ *Entrapment* means luring or pursuing someone into committing an illegal act that the individual had no original intention of committing.

√ *Espionage* is considered the act of spying on someone or something with the intent of gaining secret, personal, or classified information.

√ *Hearsay* is evidence that is not generally admissible based on the fact that it is gathered from second-hand sources and not directly tied to a first-hand specific witness.

√ The term, *enticement*, is defined as the practice or act of alluring, tempting, or attracting someone or something into doing something.

√ *Fraud* is the intentional misrepresentation of the truth in order to gain a business edge, financial profit, or something considered valuable.

√ *Ethics* are defined as a set of principles or rules that govern one's morals and actions as they apply to one's duty.

√ *Civil law* has to do with wrongdoing between people or businesses that typically result in loss or damage.

√ *Criminal law* protects society from individuals or groups that violate laws enacted by the government.

√ *Administrative law* regulates how government and agencies should conduct their affairs and business dealings.

√ The *Internet Architecture Board (IAB)* is the governor or advisor of the Internet Society.

√ If the *chain of evidence custody* is broken, or if the evidence is ever misplaced, it is likely that the evidence will prove useless to an investigation and will most likely not be admissible in a court of law.

√ The best way to fight piracy is to use only licensed software and report uses of counterfeit or illegal software usage to the original manufacturer of the software.

√ *Motive* refers to how hackers and cyber criminals are *motivated* to exploit the weakness of systems and the Internet by money, greed, and ego. As with most things in life, if a weakness exists with something, someone will be willing to exploit it.

√ *Opportunity* refers to how the Internet and weak or unprotected servers and workstations offer many weaknesses that can be exploited by hackers and would-be cyber criminals. In simple terms, the windows of opportunity are open for those who wish to exploit them.

√ *Means* refers to exploitation tools, such as "how-to" books and scripts that are publicly available to that wish of obtaining the means to commit Internet- or computer-related crimes.

√ *Cyber stalking* is defined as the continuous computerized harassment of another person.

CHAPTER SUMMARY

The obvious goal of this chapter was to sharpen your knowledge regarding computer laws, crimes, ethics, and investigation in order to better prepare you for specific Security+ exam information as it may relate to these mentioned topics. At this point, you should at the very least have a basic understanding of the following:

- Trademarks, patents, copyrights, and trade secrets.
- The important legal information and laws provided in the chapter and at the recommended Web-site references.
- Computer crime categories such as fun attacks, grudge attacks, business attacks, terrorist attacks, data diddling, and salami attacks.
- Computer and Internet ethics.
- How an investigation is carried out properly and the importance of the proper chain of evidence custody.

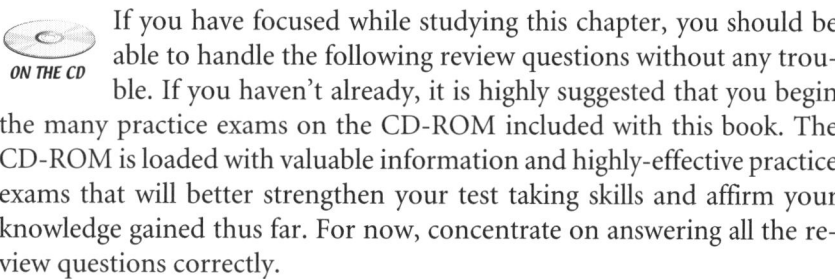

ON THE CD

If you have focused while studying this chapter, you should be able to handle the following review questions without any trouble. If you haven't already, it is highly suggested that you begin the many practice exams on the CD-ROM included with this book. The CD-ROM is loaded with valuable information and highly-effective practice exams that will better strengthen your test taking skills and affirm your knowledge gained thus far. For now, concentrate on answering all the review questions correctly.

REVIEW QUESTIONS

1. **Which of the following groups are associated with the largest total amount of computer crimes?**

   ○ A. Partners.

   ○ B. Hackers.

   ○ C. Employees.

   ○ D. Crackers.

   ○ E. Terrorists.

   Correct answer = C

   It's true; company employees represent the largest group or source of computer-related crime losses. Although business partners, hackers, and crackers often commit computer-related crimes, they do not represent the largest group.

2. **In order for computer crime related evidence to be admissible in court, it must be what?**

   ○ A. Virus scanned.

   ○ B. Encrypted.

   ○ C. Neat and orderly.

   ○ D. Relevant.

   ○ E. None of the above.

   Correct answer = D

   In order for computer crime evidence to be admissible in court, it must relevant to the specific crime that was committed. All other choices are invalid.

3. **What is the first thing you should do if an intrusion has been detected within your organization?**

   ○ A. Ensure that a proper chain of evidence custody is in place.

   ○ B. Determine the extent of damage to compromised data, systems, and networks.

   ○ C. Isolate the intruded data, system, or network.

   ○ D. Call the security guard and have all exits locked.

   ○ E. None of the above.

   Correct answer = B

   In order to properly scope and handle an intrusion, the first step is always to assess the possible damage to what has been actually compromised. Answer A looks tempting. It might be on the real exam also; unfortunately, it is not the first step. All other answers are invalid.

4. **Following the proper custody chain of evidence helps to ensure what?**

   ○ A. That other data, systems, and networks will not be compromised.

   ○ B. Admissible evidence and successful prosecution.

   ○ C. Integrity, accessibility, and confidentiality.

   ○ D. That you will not be prosecuted for being negligent.

   ○ E. None of the above.

   Correct answer = B

Following the proper custody chain of evidence helps to ensure that computer crime evidence will be admissible in court and lead to successful prosecution of guilty parties. All other answers are invalid.

5. **Which of the following involves manipulating information before or when it is entered into a system?**

   ○ A. Data diddling.
   ○ B. Data transversal.
   ○ C. Data doodling.
   ○ D. Undo care.
   ○ E. None of the above.

   Correct answer = A

   Data diddling is a computer crime that involves the changing or manipulation of data before or as the data is entered into a computing system. Typically, the data is changed back to original form after the crime has been committed. All other choices are invalid.

6. **A software company has developed new code that will change the world. It is essential that the software's company competitor does not gain access to or use this code. What is the new code considered?**

   ○ A. Patented code.
   ○ B. Copyrighted data .
   ○ C. A trade secret.
   ○ D. Really neat HTML.
   ○ E. None of the above.

   Correct answer = C

   A trade secret is proprietary company information. That secrecy is essential to the health and profitability of a company. A patent is a privilege or right of use that is specifically assigned by government to the creator. A copyright is the right to create and sell that is exclusive to the creator or owner of the copyright. Really neat HTML is nice, but it is not the answer here.

7. **Which computer crime category represents an individual whose primary goal is to access secured data and information just for the thrill of it?**

   ○ A. Terrorist attacks.
   ○ B. Grudge attacks.
   ○ C. Business attacks.
   ○ D. Fun attacks.
   ○ E. None of the above.

   Correct answer = D

   The goal of the fun attacker is of a boastful nature. "What can I get at?, "How far can I get?" and "Who can I tell how special I am for getting there?" are usually the interests of this individual or group. Terrorists utilize information systems and technology as tools to support their financial and other illegal and immoral activities. Grudge attacks are meant to cause damage to people and systems that the attacker doesn't like. Attacks on businesses and business systems are meant to reduce profits and undermine a company's reputation.

8. **Which computer-related attack crime usually goes unnoticed based on the fact that it focuses on financial gains in tiny increments?**

   ○ A. Salami attack.
   ○ B. Grudge attack.
   ○ C. Business attack.
   ○ D. Fun attack.
   ○ E. None of the above.

   Correct answer = A

   A salami attack is a computer-related attack with intention of making a financial gain using very small increments of information and money that usually go unnoticed. Grudge attacks are meant to cause damage to people and systems that the attacker doesn't like. Attacks on businesses and business systems are meant to reduce profits and undermine a company's reputation. The goal of the fun attacker is of a boastful nature. "What can I get at?," "How far can I get?," and "Who can I tell how special I am for getting there?" are usually the interests of this individual or group.

9. **What does RFC 1087 pertain to?**

   ○ A. RFC 1087 classifies all computer crimes.

   ○ B. Better defines first computer law of 1984.

   ○ C. Ethics and proper use of the Internet.

   ○ D. Refunds For Customers who have been mislead.

   ○ E. None of the above.

   Correct answer = C

   RFC (Request For Comments) 1087 pertains to ethics and proper use of the Internet. The 1986 Computer Fraud and Abuse Act better defines the first computer security law created in 1984. All other choices are invalid.

10. **Which act is concerned with the protection of information systems by deterring and obstructing terrorism?**

    ○ A. Electronic Communications Privacy Act.

    ○ B. U.S. Patriot Act.

    ○ C. U.S. Homeland Act.

    ○ D. Gramm-Leach-Bliley.

    ○ E. None of the above.

    Correct answer = B

    The U.S. Patriot Act addresses many of the growing concerns regarding the protection of information systems by deterring and obstructing terrorism is the U.S. The Electronic Communications Privacy Act prohibits eavesdropping by way of wire or oral communications without explicit permission. The Gramm-Leach-Bliley law ensures that financial institutions have an obligation to protect the privacy of their customers by implementing and supporting technical, administrative, and physical safeguards.

11. **Computer crimes are classified into two main categories. What are they?**

    ○ A. Criminal and civil computer crimes.

    ○ B. Crimes that are carried out against a computer and crimes committed using a computer.

    ○ C. Civil and administrative computer crimes.

○ D. Espionage and theft computer crimes.

○ E. None of the above.

Correct answer = B

Computer crimes are separated into in two categories. Crimes that are carried out against a computer and crimes committed using a computer. All other choices are invalid.

12. **Which of the following is considered the intentional misrepresentation of the truth in order to gain a business edge, financial profit, or something considered valuable?**

○ A. Embezzlement.

○ B. Piracy.

○ C. Fraud.

○ D. Espionage.

○ E. All of above.

Correct answer = C

Fraud is the intentional misrepresentation of the truth in order to gain a business edge, financial profit, or something considered valuable. Fraud can be also defined as trickery, deception, and lying in order to position oneself for illegal gain. Embezzlement is the illegal use of or stealing property that belongs to someone else that has been entrusted to your care. Software piracy is the illegal duplication, use, and distribution of software. Espionage is considered the act of spying on someone or something with the intent of gaining secret, personal, or classified information.

13. **What does MOM stand for?**

○ A. Motivation, Opportunity, Money.

○ B. Malicious, Open, Motivate.

○ C. Motive, Opportunity, Means.

○ D. More, Others, Money.

○ E. None of the above.

Correct answer = C

Information security specialists and several security examinations (including the Security+ exam) use the acronym, MOM, to describe the "why, when, and how" of computer crime crimes. MOM stands for Motive, Opportunity, Means. All other choices are invalid.

14. **This type of law has to do with wrongful doings between individuals or between businesses and individuals. Typically, it results in some sort of loss or damage. What type of law is it?**

    ○ A. Administrative law.

    ○ B. Gubernatorial law.

    ○ C. Legislative law.

    ○ D. Criminal law.

    ○ E. Civil law.

    ○ F. None of the above.

    Correct answer = E

    Civil law has to do with wrongful doings between individuals or between businesses and individuals. Typically, it results in some sort of loss or damage. Administrative law pertains to and regulates government agencies, organizations and offices. Criminal law protects society from individuals or groups that violate laws enacted by the government. All other choices are invalid.

15. **A disgruntled computer savvy employee (or former employee) with a score to settle is most likely to use this as a way to get revenge?**

    ○ A. Civil law.

    ○ B. Data diddling.

    ○ C. Logic bomb.

    ○ D. Chain of evidence custody.

    ○ E. MOM.

    ○ F. None of the above.

    Correct answer = C

    Grudge attacks are usually launched from remote locations using existing VPN connections or in the form of a particular type of malicious code such as a logic bomb. Civil law has to do with wrongful doings between individuals or between businesses and individuals. This would be the constructive approach to handling existing differences between employee and employer. Data diddling is a computer crime that involves the changing or manipulation of data before or as the data is entered into a computing system. The chain of evidence custody is a documented report that identifies who has custody of evidence from beginning to end. Information security specialists and several security examinations (including the Security+ exam) use the acronym, MOM, to describe the "why, when, and how" of computer crime crimes.

16. **What do most businesses use to provide surveillance of entryways and exits?**
    - ○ A. Biometric devices.
    - ○ B. CCTV.
    - ○ C. Guard dogs.
    - ○ D. Tokens.
    - ○ E. Smart cards.
    - ○ F. None of the above.

    Correct answer = B

    Most businesses monitor entryways, exits, and access to secured locations through the use of Closed-Circuit Television (CCTV). Biometric devices, tokens, and smart cards are primarily used as authentication mechanisms. Guard dogs are used as physical deterrents or physical barriers.

17. **This law was enacted for the further protection of nonpublic personal information. Its main focus is to ensure that financial institutions have an obligation to protect the privacy of their customers by implementing and supporting technical, administrative, and physical safeguards. What is this law?**
    - ○ A. Electronic Communications Privacy Act.
    - ○ B. U.S. Patriot Act.
    - ○ C. U.S. Homeland Act.
    - ○ D. Gramm-Leach-Bliley Law.
    - ○ E. None of the above.

    Correct answer = D

    The Gramm-Leach-Bliley law ensures that financial institutions have an obligation to protect the privacy of their customers by implementing and supporting technical, administrative, and physical safeguards. The Electronic Communications Privacy Act prohibits eavesdropping by way of wire or oral communications without explicit permission. The U.S. Patriot Act addresses many of the growing concerns regarding the protection of information systems by deterring and obstructing terrorism is the U.S. Choice C and E are invalid selections.

REFERENCES

http://www.complaw.com is hosted by CompLaw. It has a marvelous computer law library and posts very recent news regarding businesses and technical companies that are making or breaking computer laws.

http://www.findlaw.com is hosted by FindLaw, is a great research tool that can be used to find just about anything related to computer laws, legal representation, and case history.

http://www.usdoj.gov/criminal/cybercrime/cclaws.html is offered by the Department of Justice. It contains a wealth of information regarding computer crime, laws, and policies.

http://www.cdt.org/security/010911response.shtml is offered by the Center for Democracy and Technology Web site has a wealth of information related to the U.S.A. Patriot Act as well as other anti-terrorist information and measures.

http://www.techtv.com/cybercrime/aboutus/story/0,23008,3339221,00.html is offered by TechTV Inc. It contains some excellent information that can help you become better educated and assist you with avoiding computer fraud and other threats.

http://www.brook.edu/its/cei/cei_hp.htm is offered by the Brookings Institution and describes computer ethics is great detail. It also contains the Ten Commandments of Computer Ethics. © 1991. Computer Ethics Institute.

http://www.cis.ohio-state.edu/cgi-bin/rfc/rfc1087.html describes the IAB's (Internet Activities Board) Request for Comments: 1087 concerning Ethics and The Internet.

http://www.cerias.purdue.edu/coast/ is offered by Purdue University and is the home of COAST. It is one of the largest security research groups in the world. They provide a wealth of valuable information regarding computer investigations.

CHAPTER 9

VIRUSES, WORMS, TROJANS, AND PROTECTION

IN THIS CHAPTER

- Introduction
- Malicious Code
- Specific Computer Viruses
- Antivirus Solutions
- Operating Systems Security
- Test Tips
- Chapter Summary
- Review Questions
- References
- Introduction

You might be asking yourself, "Why is there an Introduction in the last chapter of this book?" Here is the answer. If you have focused and really dedicated yourself to the first eight chapters of this book, it is likely that you already posses the skills and tools necessary to pass the current Security+ examination. You are now being introduced to the chapter in this book, because it is a book within itself. It might not be as lengthy as some of the previous chapters in this book. However, it will serve as your silver bullet in the exam room. This chapter is specially designed to give you that extra edge that will increase your chances of scoring higher on the exam.

Note that some of the terms that you will see in this chapter have been described to you briefly in earlier chapters—for examples, virus, logic bomb, and Trojan horse. This method has been implemented purposefully as a *staircase* approach throughout this book in order to prepare you for the overall understanding that you will achieve by the end of this chapter.

After looking over the CompTIA Security+ exam objectives, most likely, you have come to the conclusion that you will have to be familiar with certain types of malicious code, viruses, and virus categories as well as protecting certain operating systems in order to pass the test. However, which viruses should you concentrate on? Which operating systems will be targeted on the exam? When this book was written there were over 65,000 known viruses. Are you familiar with all 65,000? As far as operating systems go, there are four different operating systems from four different venders that CompTIA will most likely target on the examination. After all, the exam is "vendor neutral."

No need to fear if you spend a good portion of your life studying for, fine tuning, and taking certification exams from the major vendors such as Microsoft and CompTIA. You will begin to notice a very evident pattern that develops with most if not all of their exams. They seem to take the top ten or 20 most important topics regarding particular subject matter and design two or three questions around each of those individual topics. The questions, although they appear to be unrelated, typically all lead to the same place and the same predictable answers. It's in the way they ask you the questions that the correct answer can be found. In other words, the questions are designed to take you away from where you should be focusing by throwing in tempting garbage that has little or nothing to do with the correct answer. Your job is to separate the truth from the garbage.

It is likely that you will need to know the top one or two viruses from each of the virus variation types over the last several years that have affected the planet in a negative way. We will focus on them and more. It is also likely that you will be asked general questions relating to basic security functionality within Microsoft Windows operating systems and UNIX/Linux. It is recommended that you focus on these particular operating systems for they are the most commonly used, administrated, and often-targeted operating systems by hackers. CompTIA knows this. As a result, it is highly likely that the majority of questions you face will target these two operating systems. Increase your odds of passing by focusing there.

It is likely that the current Security+ exam will not target specific operating systems with specific detailed operating systems questions by using particular screen shot diagrams of the inter workings of the operating systems. In simple terms, you might see general operating systems questions similar to the following, as opposed to questions that ask how you would specifically carry out a particular task:

What is the Administrator account called using Linux?

Where is the administrative password stored in Linux by default?

What is the archival (backup) command called in Linux?

What account should you rename using Microsoft Windows?

 This exam leans more towards generalizations than specific technical operating system tasks. It is more likely that you will be presented with general routing tables than specific operating systems screenshots or diagrams.

Update your virus definitions, do a full system scan, enable your stateful inspection firewall, and fasten your seat belts. Here we go.

MALICIOUS CODE

Malicious code is defined as software that has the ability to carry out or perform certain unauthorized functions that disrupt or interfere with normal operations on a computing system. There are several general types of malicious code that you must be familiar with for the Security+ exam. You must know the differences and similarities between them. After becoming familiar with these classifications of code and other descriptions, we will move on to the more specific identification of the viruses CompTIA is most likely to expect you to know. Pay attention here! This information is likely to appear on the exam and is very critical to your success.

VIRUSES

A *virus* is a program or specific piece of code is that is designed, when executed, to duplicate itself and/or spread itself to other areas of a system or other systems in a networked environment. In general terms, a virus will replicate itself until it uses up all available system resources such as memory or hard drive space. The result of an undetected virus that has infected a system and been successful at achieving its goal is a system that simply will not function. In most cases, the system will end up unavailable to other systems or result in a Denial of Service.

WORMS

A *worm* is a type of virus that gets its name from its inherent ability to spread itself to other networked systems, remain resident in memory, and keep in contact with other segmented pieces of itself until triggered by a certain event to duplicate and spread itself.

Most worm viruses reside in memory, unattached to files, and when triggered, will reproduce themselves until available resources are exhausted. A worm is a self-contained unit or program that is typically spread through e-mail attachments and network connections such as drive mappings.

 It is very important to understand that a worm is a type of virus that can replicate itself. However, worms do not attach to other programs. In other words, worms are not carried by or attached to hosting files.

TROJAN HORSES

A *Trojan horse* is a program that appears on the outside to be harmless. It masquerades as an apparent, nondestructive, and innocent application, program, or message. Most Trojan horses carry very dangerous payloads that are often times highly destructive to networks and systems.

Most Trojan horses are hidden in Internet attachments that are often times distributed with e-mail and in the form of jokes, love letters, and misguiding advertisements. One of the most important facts to understand regarding Trojan horses is that they do not replicate or copy themselves. They require actions on the behalf of the user to activate and deliver their dangerous contents. This type of action can be the opening of an attachment or the running of an application.

 It is very important to understand that while worms and viruses duplicate themselves, Trojans do not.

LOGIC BOMBS

As you may recall from your study of Chapters 7 and 8, *logic bombs* are considered malicious code that are inserted into a operating system or application that is set to "explode" or go off when triggered by a certain time, date, or event. In simple terms, a logic bomb can be a virus or Trojan horse that activates when certain conditions are met.

Some logic-bomb code is very tricky and hard to detect as it just sits there and waits for certain criteria to be met before activation. Some antiviral scanners have the ability to detect logic bomb code. For the best possible protection from logic bombs, it is suggested that each computer system have individual antivirus software protection, network screening controls be in place, real-time virus protection be enabled, and all e-mail scanning functions are enabled.

 Logic bombs are often left behind by disgruntled former employees with a grudge or a score to settle. In most cases, these bomb planters are technically savvy programmers, developers, or network administrators.

FILE INFECTORS

File infectors are viruses that attach themselves to files that can be executed, typically. When the program or application is executed, so is the virus attached to it. The types of viruses usually associate themselves with EXE and COM files.

SYSTEM OR BOOT INFECTORS

System or *boot infectors* are viruses that are commonly known to attach themselves to and damage system files such as a hard drives Master Boot Record (MBR) or the boot sector on a floppy disk. Simply put, if infected these types of viruses will be triggered when you system is booted from you hard drive or floppy disk. You will most likely end up with a trashed hard drive if you are infected with this older type of virus. You might have to do a low-level format of your hard disk and reload the operating system.

MACRO VIRUSES

A *macro virus* is a virus that utilizes another application, such as Microsoft Word or Excel, macro code, or programming language to be distributed. Macro viruses are the most common type of viruses. They are considered to be a minimal threat based on the fact they do not infect a system's boot sector or actually infect other programs. Most macro type viruses are designed to insert numbers, characters, words, or phrases into documents or spreadsheets. If you have a good antivirus solution with updated definition files and real time protection activated, most common macro viruses will be detected and cleaned.

STEALTH VIRUSES

A virus with *stealth* characteristics will hide itself and send bogus responses back to an antiviral software packages scan request in order to avoid detection. In other words, stealth viruses "lie" to antivirus packages by making sure that certain files are in working order. When a stealth virus attacks, typically, it will make a good copy of a file that it attacks. When a scan request is initiated, the virus will send the good copy of the file to scan package. The original Brain virus was a stealth type virus that would infect a hard drive's boot sector. It would remain memory resident, masquerade itself as a good file and fool the operating system and scanners into believing everything was hunky dory.

A stealth virus can be passed to many systems and files in a networked environment continuously tricking port sniffers and antivirus software scanners. Stealth viruses can infect boot sectors as well as proliferate with the execution of a program or with the simple event of someone opening a folder or file. Typically, stealth viruses receive very high virus threat ratings (these are described shortly).

POLYMORPHIC VIRUSES

A *polymorphic* virus possesses the ability to change its own internal code and byte structure as it is being duplicated. This ability to change itself and appear to take on multiple forms of existence is referred to as *polymorphism*. Viruses that have polymorphic qualities are very difficult to detect by signature scanning antivirus software packages.

BLENDED THREATS

Blended threats are viruses that combine the most lethal characteristics of viruses, worms, and Trojans in order to cause mass destruction and wreak total havoc on a targeted network or system. Blended threats can spread through a network or multiple networks very quickly. They are becoming popular tools among those who wish to cause mass damage. Weak or unprotected networks and systems stand little chance of survival against blended attacks. Blended threats will usually include the following characteristics:

- They will spread automatically by continuously scanning the Internet for Web servers with open or vulnerable TCP/IP ports.

- They will usually insert Trojan or logic-bomb code in targeted servers for timed attacks.
- They will create network shares, change account privileges, and utilize existing network mappings as ways to propagate within an infected network.

VIRUS VARIANTS

Variants are new viruses or *virus strains* that take code and sometimes modify the code of existing well-known viruses. Existing well-known viruses are considered somewhat easy to predict and protect against with properly updated and configured quality anti virus software. However, with new strains of viruses constantly being developed and introduced into our electronic world, we must continually remain diligent in order to protect or corporate as well as private (home) electronic jewels. In other words, the major antivirus software manufacturers need to plan for and anticipate the development of virus variations by continuing to create and improve upon virus packages and definitions that can identify malicious core code. Corporate management needs to budget and purchase quality antivirus software and hardware and network security folks must properly configure and manage the gifts they might be blessed with.

Virus variants and their payloads (these will be described shortly) can be compared with terrorist attacks. It is not obvious when, where, or how they may occur but it is quite evident that we must anticipate and prepare for them in order to detour or minimize their effects.

RETROVIRUSES

A *retrovirus* is designed to attack antivirus software programs first with the intent of passing through undetected. It is not likely that the current Security+ exam will target this definition. However, it is important as a computer security professional that you are aware of what a retrovirus is.

VIRUSES "IN THE WILD"

Viruses that are "in the wild" exist and spread on systems and in networks that are commonly used on a daily basis. Viruses that are wild are also viruses that exist outside of contained and controlled environments, such as registered scientific antivirus research systems. An organization known as The Wild List Organization International maintains a constantly up-to-date list of what viruses that are "in the wild." You can learn more about The Wild List Organization International and the "in the wild" virus list by visiting the following Internet site: *http://www.wildlist.org/*.

ZOO VIRUSES

A *Zoo* virus threat is a threat that only exists in contained, controlled antivirus labs. A *Zoo threat*, or virus that resides in the Zoo laboratory, is not considered to be a threat to normal everyday systems and networks. In other words, it is not "in the wild." Typically, Zoo viruses are used to test the responsiveness of a particular software product that is in development. In other words, researchers will see how a program reacts when viruses are introduced.

VIRUS PAYLOAD

The actual action that a virus carries out is called the virus's *payload*, which can be either malicious or harmless. Malicious payload could be the reformatting of your hard drive, deletion of certain files, or attempts to access confidential information such as a back account or credit card information that is stored on a system. Harmless payload can be described as a pop-up message that tells you how smart you are or displays some other form of advertisement. Many virus payloads are triggered by a certain date, time, or event. This date, time, or event is known as the payload *trigger*.

VIRUS THREAT RATING

A virus *threat* or *risk* rating is a calculated value that represents the possible level of severity or threat that an identified virus or piece of malicious code represents to a computer system. A virus's risk rating is calculated with several factors in mind. Most often factors such as the number of attacks reported, the ability for a virus to replicate itself and spread, as well as the severity and possible damage that a virus's payload can cause, are used to calculate a virus's severity of threat rating. The major antivirus solution providers such as Symantec and McAfee post the most common, as well as past reported, viruses and their potential threat or risk ratings on their Internet sites.

MALWARE

Malware is a term that is used loosely to describe unwanted malicious software and other software that is just plain unwelcome. Malware can be viruses, Trojans, or worms. Simply put, Malware is shorthand for malicious code. It is something that produces unwanted, unexpected results.

SPYWARE

Spyware is a program or piece of software that remains hidden on a system that monitors and logs another system's activities. There are many free as well as commercial spyware programs that you can get which will allow you to secretly record, capture, and store Instant Messaging conversations, e-mails, and other related data. There are also particular spyware-removal software products available that will assist you with the removal of spyware software that might be hidden in your system or somewhere in the labyrinth of your corporate enterprise.

TERMINOLOGY

It is important that you are familiar with virus naming terminology in order to assist you with the identification of certain virus types. If you surf the Internet for virus names, specifically the sites provided by the major virus software manufactures, you would notice that many of the computer viruses listed include *.dr, .enc, @m,* or *@mm,* in their names. For instance, take the virus name *W32.Nimda.A@mm.* The *@mm* in this virus's name signifies that this particular virus is a mass mailing worm virus. Please familiarize yourself with the following basic terminology for the exam:

- **.dr:** This represents files that are considered to be *dropper* files. These are programs that drop a virus or worm onto a victim's computer system.
- **.enc:** This refers to a file that has been encrypted or encoded. Viruses commonly use these types of files to hide themselves.
- **@m:** This refers to *mailer* worms, which are viruses that attach themselves to mail which the victim sends in order to spread.
- **@mm:** This refers to *mass-mailer* worms, which are viruses that attach themselves to malicious mails and are sent automatically to contacts in an address book.

There are many variations of viruses, worms, and Trojans. Next, we will get specific with our study of malicious code in order to prepare you for the worst.

SPECIFIC COMPUTER VIRUSES

It was stated in the introduction to this chapter that CompTIA includes the subject of viruses in its domain objectives. It was also stated that there are over 65,000 reported or known viruses and variations. However, CompTIA does not specify which specific viruses it will target in the Security+ exam.

You'd better believe the exam is going to ask you to identify specific viruses. With this in mind, it is important that you are familiar with as many specific viruses, worms, Trojans, and other malicious code. If you focused earlier with your study of malicious code, you should be more than ready to remember and associate the following specific viruses. Pay close attention here. This just might get you a passing grade on the real examination.

BACKDOOR.SUBSEVEN

Backdoor.Subseven is very similar to the remote administration tool threats Back Orifice and NetBus (which will be described shortly). It is considered a Trojan horse that allows a computer to be controlled remotely from another location. In other words, it allows undetected, unauthorized access to your system from within a network or over the Internet. The Backdoor.Subseven virus and some of its known variants—Backdoor.Sub-Seven.1_7, Backdoor-G, Backdoor.Trojan, and Sub7 are most commonly distributed through e-mail attachments and instant messaging file and program transfers.

Once executed, the virus will basically add itself, as well as other related executable files, into the system folder within your operating system. It also changes system registry values, which will allow your system to eventually be controlled. The affected system becomes the "server" that can then be controlled by a remote "client."

If you have been infected with the Backdoor.Subseven virus, some or all of the following and much more might result:

- Your system can be remotely shut down and or restarted.
- Confidential information can be browsed, deleted, or obtained.
- Programs can be deleted or modified.
- System files can be manipulated.

BACK ORIFICE (A.K.A. AS BO TROJAN)

The *Back Orifice Trojan horse* is a program that is similar in nature to Net-Bus (described next) that allows remote access to a computer system after a server application program has been executed on the remote or targeted computer. After the server piece has been executed on the target system, the remote system (client) can do just about anything they please to the infected system. The insidious thing about this is that the program can do its bidding undetected.

NETBUS (A.K.A. BACKDOOR.NETBUS)

NetBus is a remote administration Trojan horse type program similar to Back Orifice and Backdoor.Subseven that must first be executed on a system by a user in order to be installed. A system affected by NetBus or its variants can expect many of the same results as described with Backdoor.Subseven and Back Orifice. However, there are a few distinctions that exist with NetBus that make it a remote administration Trojan horse of choice for many hackers. It allows a remote user to pop a system's CD tray in and out as well as manipulate mouse buttons and pointers. It should also be noted that there are several versions of NetBus. Unlike earlier versions of this remote control program, NetBus Pro Version 2.1 has been designed not to hide itself as to advertise that is a legitimate, controlled, remote administration tool.

If you research the Back Orifice and NetBus, you will most likely become very confused by the various descriptions and other explanations of these threats. Some sources simply describe them as "authored remote administration tools" while others, describe them as tools that are modified for malicious purposes. This type of subjective matter and confusion would be perfect ground for the trickery typically found in Microsoft certification examinations. Fortunately, the Security+ is a CompTIA certification. CompTIA will most likely be concerned that you know these are modified remote administration programs or tools that can are used for unauthorized, illegal and malicious purposes. For your own sanity and our specific Security+ exam focus, know that the Backdoor.Subseven, Back Orifice, and NetBus are considered to be Trojan horse remote administrative threats that exist in the wild.

CHERNOBYL

The *Chernobyl* virus, also named *W98.CIH* or just *CIH*, was named after the Chernobyl nuclear disaster and its author (Chen Ing-hau). It is an older *space filler* virus that mainly targets earlier versions of Microsoft Windows such as Windows 95 and Windows 98 operating systems. It was a devastating virus that would fill up all free space areas on a hard drive, making it very difficult for antiviral software to run, and was capable of wiping out all data in an infected computer system. This virus was triggered on April 26th, which coincides with the date of the Chernobyl nuclear disaster in Russia on April 26th, 1986. This virus is still considered to exist in the wild. Fortunately, it can be identified and controlled with current managed antivirus protection and updates.

ILOVEYOU

In May of 2000, The ILOVEYOU worm was released on the world. ILOVEYOU is a self-propagating worm that is included as an e-mail attachment to an e-mail entitled *ILOVEYOU*, which typically would be addressed to a targeted address from a friend, loved one, or associate. Once the attachment included with the e-mail is opened, the worm infects files with extensions such as *.vbs, .vbe, .com, .jpg, .jpeg, .gif, .doc, .hta, .mp3, .wav, .txt, .bat, .htm,* and *.html,* just to name a few. The worm then sends itself to all contacts in a targeted systems local Microsoft Outlook address book. This worm was one of the fastest spreading worms to date based on its marveled social engineering techniques; ability to be proliferated to all contacts within a contact list (the Melissa virus used the very first 50 contacts listed in a local address book); and its ability to spread through mapped network resources quickly.

MELISSA

The *Melissa (W97M.Melissa.A) virus* is a macro virus that spreads very quickly when its payload is released or executed. The Melissa virus is distributed as an e-mail attachment, most often named LIST.DOC, whose mail subject title reads, "Important Message from [the name of someone]" and body text that reads, "Here is that document you asked for. . . don't show anyone else ;-)". When a person opens the attachment, the virus infects the targeted system, corrupts certain files and safety features associated with Microsoft Word, and e-mails itself to the first 50 contacts in the system's local Microsoft Outlook address book.

W32.KRIZ

The *W32.Kriz* virus is a virus similar to the Chernobyl virus, which typically resides in computer systems memory. Its payload attempts to *flash* or erase a computer system's BIOS as well as erase files that reside on the infected computers hard disk, floppy disk, and all associated network mapped drives. The virus creates a bogus *Kernel32.dll* file and overwrites the existing known good *Kernel32.dll*. When the programs make application program interface calls, the bad *Kernel32.dll* will infect them. If the W32.Kriz virus has infected your system, you will most likely end up spending your Christmas holiday reformatting your hard drive as well as reloading your operating system. If this virus resides in a system, it is triggered to release itself on December 25 of any given year.

W32.NIMDA.A@MM

W32.Nimda.A@mm is a mass-mailing worm that targets the weaknesses of vulnerable, unpatched Microsoft IIS (Internet Information Server) Web servers. W32.Nimda.A@mm proliferates through e-mail attachments and uses the Unicode Web Traversal exploit. Once it has infected an unpatched server, the server acts as a host or catalyst, if you will, which allows the worm to search through mapped network shares for other weak IIS servers within the network. Nimda affects local system as well as remote network shares and files. Here are two interesting facts: the virus' name comes from the reversed spelling of *admin*. During the infection process, the local guest account is created with administrator privileges. This allows the worm to do its bidding on a local system.

When Nimda was released, it affected thousands of Web servers across the world. The author of this book personally contributed hundreds of hours successfully saving many infected IIS servers across a corporate network from destruction by this worm. Although the CompTIA Security+ exam and this book are considered "vendor neutral," the author of this book is somewhat partial and would like to thank technical support at Symantec Corporations, "the world leader in Internet security technology," for their quick response and assistance with fighting this worm. In the author's opinion, Symantec Corporation was the first antivirus software provider to produce an effective solution (patches and instructions) for this worm that saved a very important company a lot of money and busy author a lot of time and frustration.

There are several variations of Nimda that you should acquaint yourself with. It is highly recommended that you visit the Symantec Security Response Web site and educate yourself with the Nimda variations as well as other viruses, worms, and Trojans.

 It is likely that the Security+ exam will expect you to know which major types of virus codes NetBus, Nimda, Brain, Backdoor7, and BackOrifice are each associated with. In other words, what types of viruses are they? Know them well!

If you are interested in getting more acquainted with some of thousands of viruses that exist, Symantec has quite an extensive virus encyclopedia that is available to the public at *http://securityresponse.symantec.com/avcenter/vinfodb.html/*.

W32.KLEZ.A@MM

W32.Klez.A@mm is a mass-mailing e-mail worm threat that exploits known weaknesses associated with Microsoft Outlook Express and Microsoft Outlook. (Remember what the *mm* means?) This virus is very tricky; it spreads itself to local as well as mapped network drives when it is opened or viewed. Basically, the virus will zero out all infected files causing them to be useless. The virus's payload is date specific. This means that when a certain system date is reached, the payload will be executed. The virus in known to release itself when the system date reaches the 13th of January, March, May, July, September, and November. If you have a quality antivirus product that is properly configured with updated virus definitions you should be well protected from this threat. If you do not, and have been infected by this virus, it is highly recommend that you visit the Symantec Security Response page at *http://securityresponse.symantec.com/* and acquire the necessary removal tool for the worm.

 Mass-mailing worms pose very serious threats to your system. They are very common today and are likely to be targeted on the exam. Know the most popular ones well!

Next, we will discuss the importance of managing and maintaining antivirus solutions properly.

ANTIVIRUS SOLUTIONS

Whether you have a stand-alone computer system, home network, or are fortunate enough to have the responsibility of protecting an enterprise network, it is critical that you properly implement and maintain a good solid antivirus solution. Insidious hackers continue to develop new tools and techniques with the intentions of demolishing your home or corporate information. In order to give your important enterprise platforms and data a fighting chance, you must take the time to educate yourself with antiviral products that provide enterprise level support and centralized management.

It is very common today to find home computer systems as well as enterprise business systems that do have quality antiviral products installed from major manufactures such as Symantec or McAFee. However, it also very common to find many of these installations improperly configured and mismanaged. Typically, this results in systems that appear to protected, when in reality, they are sitting ducks for viruses. There are several reasons for this. It is very common to find systems that do not have their virus definition files up-to-date. This results in a lack of protection against new viruses and variations of older viruses.

It is crucial to the welfare of your systems and data that you keep your virus definitions files up-to-date. Most quality antivirus packages offer the ability to have your system retrieve virus definitions automatically from the manufacturer's Web site on a scheduled basis. It is highly recommended that you schedule auto updates to occur as frequently as possible. Symantec enterprise antivirus software as well as their personal antivirus editions call this LiveUpdate.

By default, most of the major antivirus packages enable *file system real-time protection*. This option allows the antivirus package to continually scan files that are accessed, created, or modified at all times. Real-time protection generally runs as a service and can at times hamper your systems performance. However, it is well worth the trade off for the protection it offers. It is common to find file system, real-time protection disabled on systems. This can result in systems being wide open to virus attacks. You might not even know you have a virus until your system performs a system virus scan. This brings up another point: if you do not schedule full system scans to occur on a frequent basis, you might not even know you have a virus until it's too late.

The point here is that just having antivirus software installed does not mean that you are properly protected. Education, training, and proper daily management of antivirus products, particularly in a networked environment with enterprise support, are a must.

Next, we will focus on antivirus practices that should be implemented in order to better increase the chances of survival of your home or enterprise systems or networks should they (and they will) come under virus attack.

ANTIVIRUS PRACTICES

As stated earlier, a combination of education, training, and management policies with antivirus products and practices is essential to the survival and welfare of computer systems and networks.

The following basic guidelines should be followed when using and managing antivirus software at home or in a business enterprise:

- Install, update, and maintain reputable quality antivirus software to servers and workstations. This includes setting up daily antivirus definition updates, enabling real-time protection, setting up schedules scans of all system drives, and enabling e-mail and attachment scanning.

- All users of computer systems (at home or in the workplace) should be educated/alerted when particular virus attacks occur or are expected to occur. Symantec Corporation does a great job updating their Web site with virus threats that are anticipated. This information can prove invaluable to administrators who need to apply particular patches or make updates in preparation for new or anticipated variations of viruses.
- Teach all users at home or in the workplace that opening e-mail attachments as well as Instant Messaging attachments might be detrimental to your system's life span.
- In a business environment, ensure that your corporate antivirus business policy is up-to-date and accurate. Ensure that new and existing employees sign an addendum that states they are familiar with the company's policy regarding computer usage as well as virus policy and procedures.

OPERATING SYSTEMS SECURITY

The information contained in this section is meant as a basic overview of some of the basic administrative information, as well as tasks, that should be performed in order to harden the operating systems that will be targeted on the exam. It is very likely that you will encounter questions related to these operating systems' fundamentals on the exam. Please keep in mind, this overview is by no means meant to be a detailed "how-to" administrative reference. Instead, it isolates the information that is most likely to appear on the exam.

Throughout this book, you have been introduced to tools, resources, and practices used to harden or better ensure the integrity of your systems and overall network. No matter what operating systems you are protecting, the following well-known practices should be implemented:

- **Remember to stop or remove any unneeded services.** Many default operating system installs include and run services such as FTP, Web services, and Telnet. These services require hardening and continuous patches. Many operating system services provide weakness and leave systems open to attack. Regardless of the operating systems you are using, consider removing or stopping any unneeded services.

- **Keep your service packs and patches up-to-date, no matter what operating systems you are running.** Variants of viruses and blended threats are developed continuously to take advantage of operating system weaknesses. Operating system software vendors recognize this and develop patches to secure inherent weaknesses.

- **Set up e-mail servers to scan e-mail attachments.** Also configure e-mail servers to block attachments with extensions that are known threats. These usually include attachments with extensions such as *.bat*, *.vbs*, *.exe*, *.hta*, and *.scr*.

- **Isolate any system that is known to be infected—this is important in order to reduce the chances of other systems being infected.** Disconnect the infected system from the network. Then follow your corporate or business policy regarding the extraction of evidence (forensics).

BASIC WINDOWS SECURITY

As you are most likely aware, many versions of Microsoft Windows exist in the world today in both homes and in business. The current Security+ examination is most likely to focus on the underlying administrative and basic security principles generally associated with Microsoft Windows operating systems. Throughout this book, you have been introduced to administrative principles that apply in general to most Microsoft operating systems. This section is in place to sharpen that knowledge.

For our specific security-related study focus and the obvious fact that Microsoft operating systems such as Microsoft Windows NT and Windows 2000 are currently the most widely used Enterprise level operating systems, we shall target their administrative security basics first.

The following administrative actions (hardening) should be considered after installing Microsoft Windows operating systems. If you are already running a Windows operating system, you should also consider the following recommendations:

Rename the Administrator Account

The most powerful account included with most Microsoft Windows versions is called the Administrator account. Hackers will often target this account with password breaking programs. It is good practice to rename the Administrator account to something less obvious, in order to reduce possible threats and access to this powerful account.

It is likely that you will encounter a question on the exam that asks you which account you should consider renaming in order to protect a Windows operating system better.

Verify That the Windows Guest Account Is Disabled

The Windows Guest account is disabled by default with Windows NT, 2000, and XP. It is important that you verify that this account is disabled and remains disabled. If the Guest account is enabled, it can be used to access shared resources without entering a password for authentication. This is a major security issue. As mentioned earlier in the chapter, Nimda enables the Guest account and takes advantage of its inherent weaknesses. For the exam, it is sufficient to know that the Guest account should be and remain disabled.

The following miscellaneous administrative functions should also be considered when using Microsoft Windows operating systems.

- Remove full control permissions for the Everyone group. For even stronger security, remove the Everyone group from directories altogether.
- Enable auditing and logging.
- Remove access to default administrative shares.
- Use strong passwords that use a combination of letters, numbers, and symbols.
- Allow administrative access to systems by local sign-on only.

BASIC UNIX/LINUX SECURITY

If you're coming from a Microsoft point-of-view, UNIX-based operating systems can seem quite foreign. File systems, access control methods, user accounts, and administrative tasks are designed to work in conjunction with the UNIX kernel—the central component of this OS. Security issues within UNIX-based operating systems are handled in a different manner than their Windows counterparts and we'll discuss a few items that might show up on the exam.

Linux, arguably the most popular UNIX variant, is actually based on the Linux kernel, which resembles UNIX in many ways but was written from scratch. Just be aware that although Linux and UNIX are often used within the same sentence, they are different animals. With that in mind, let's take a look at some UNIX/Linux security issues, as well as a few basic concepts that separate UNIX-based OSes from the Microsoft variety.

Although there are *window managers* that offer a graphical interface with the underlying OS, UNIX is traditionally managed through a command-line interface. Working with commands requires exacting knowledge of command syntax and the file system itself. In Linux, commands can even be case-sensitive. The most apparent difference in a UNIX environment is the structure of the file system. Instead of drive letters, UNIX and its cousins use a hierarchical set of directories that stem from the top-level directory called *root*. Root is represented by a slash (/), so a full path in UNIX would be displayed as */etc/testfile*. The first slash represents root and what follows is the directory etc where the file *testfile* is stored. Hidden directories in UNIX are represented by a dot (.) that precedes the directory name. For example, the directory */home/.secret* would be hidden to casual viewers. Following are a few important directories to be aware of in the UNIX environment.

 You must know the following UNIX directories for the Security+ exam.

- */home*: This directory contains subdirectories named after each user of the system. Personal files are stored here and access is typically limited to *root* and the owner of the directory.
- */etc*: This directory contains system configuration and administration files.
- */bin*: Named after its binary contents, this directory holds frequently used executables.
- */dev*: This area contains files that point to the physical devices attached to the system.
- */tmp*: As the name implies, this directory stores temporary files and is referred to as the *scratch* area.

Keep in mind that *root* is also the name of the administrative user account in UNIX and Linux. The most important step in the installation of a Linux system is setting the password for *root*, and as always, stronger passwords are better. The *root* account has system-wide access, and running certain applications as root can cause great harm to the system. Because of the dangers of using the root account, limited-access user accounts can be configured to allow system access to novice users. This is especially helpful when using one of the graphical interfaces to operate a UNIX system. Because these window managers can make changes to the system on their own, their use is not recommended while logged in as *root*.

One of the greatest challenges in making Linux user friendly has been the creation of a standardized package management system. In Windows, packaged .*exe* files work great for installing programs because Microsoft has retained a somewhat consistent method for placing system files. However, the many flavors of UNIX and Linux can vary greatly with regard to the placement of system files. Red Hat has developed one such package management system for Linux called *RPM*. Although it has its quirks, the RPM system has improved the way that applications are installed in Linux. Just like .*exe* files, .*rpm* files present security concerns. Even if a user is not logged in as *root*, running an .*rpm* file might require root privileges. If a malicious .rpm file is allowed root access, it might be able to damage the system.

You might be asked to identify the location of specific system files in the UNIX hierarchy. The following list points out the most significant files and their place in the file system:

- **Unencrypted passwords:** These are stored in the */etc/passwd* file.
- **Encrypted passwords:** These are stored in the */etc/shadow* file, if shadow passwords are set. If other forms of security such as MD5, DES, PGP Keyserver, or LDAP were in use, then the case would be different.
- **Environment variables:** These are stored in the hidden */home/~username/.cshrc* file.
- **Common ports and services:** This data is found in the */etc/services* file. For example, this file tells the system that the Telnet service runs on the TCP port 23.
- **Login settings:** These are stored in the hidden */home/~username/.login* file. The commands in the .*login* file are executed immediately upon logging into the system.
- **List of IP addresses mapped to hostnames:** These are stored in the */etc/hosts* file.
- **List of hosts allowed to access the Internet:** These are stored in the */etc/hosts.allow* file.
- **List of hosts prohibited from accessing the Internet:** These are stored in the */etc/hosts.deny* file.

Now, let's look at a few basic commands in the UNIX/Linux world:

- **tar:** This command is similar to the DOS *copy* command and is used also to make backups.

- **ls:** This command lists the contents of a directory. It is the DOS equivalent is *dir*.
- **rm:** This command deletes files.
- **passwd:** This command is used to create or change user passwords.
- **chmod:** This command changes the access permissions (or mode) of a file. For example, you might *chmod* a file from read-only to read-write-execute.
- **gzip:** This command compresses files. To uncompress, the command *gunzip* is used.
- **pwd:** This command displays the complete path of the current directory. Literally, it stands for *print working directory*. Don't confuse it with *passwd*.
- **rlogin:** This command is used to initiate remote access sessions.
- **kill:** This command terminates application processes.
- **make:** This command is used to install programs; it requires *root* privilege.
- **su:** The *superuser* command is executed to allow temporary *root* privileges and requires the *root* password. Because a non-*root* user can execute this command, the *root* password must be protected vigorously.

Now, let's take a look at a few tactics that prevent security breaches in a UNIX/Linux system. The most critical security concerns are related to the *root* user account. Operating as *root* can expose a system to Trojans and other maladies. Obviously, leaving an unattended system logged in as *root* is also a bad idea. The remote login capabilities for the *root* account should be disabled. The file, */etc/security*, contains data regarding where root can log in from. Any undesired points of login should be commented out of this file. As with other OSes, always disable unused ports, services, executables, and protocols. Disable unused or dormant user accounts. Use encryption whenever possible and employ SSL and SSH to protect services and protocols. Make sure that the shadow password file is in use. If it's installed, disable autologin.

TEST TIPS

As stated throughout this book, these Test Tips are useful tools that you should skim over just before you take the exam. Use them as a final cram for the exam.

- √ A *virus* will replicate itself until it uses up all available system resources such as memory or hard drive space.

- √ *Spyware* is a program or piece of software that resides hidden on a system that monitors and logs the systems or another systems activities.

- √ A *worm* is a type of virus that can replicate itself. However, worms do not attach to other programs.

- √ Worms and viruses duplicate themselves, Trojans do not.

- √ *Malware* is shorthand for malicious code. It is something that produces unwanted, unexpected results. It is a virus, Trojan horse, or worm.

- √ *System* or *boot infectors* are older viruses that damage system files such as hard drives, the Master Boot Record (MBR), or the boot sector on a floppy disk.

- √ *Variants* are new viruses or *virus strains* that sometimes modify the code of existing well-known viruses.

- √ Most *macro*-type viruses are designed to insert numbers, characters, words, or phrases into documents or spreadsheets.

- √ Viruses that are *in the wild* exist outside of controller virus research labs. Viruses that exist in these labs are known to be *Zoo* viruses.

- √ The actual action that a virus carries out is called the virus's payload.

- √ A virus *threat* or *risk* rating is a calculated value that represents the possible level of severity or threat of a specific virus.

- √ Most *Trojan horses* are hidden in Internet attachments that are often times distributed with e-mail and in the form of jokes, love letters, and misguiding advertisements.

- √ A *logic bomb* can be a computer virus or Trojan horse that activates when certain conditions are met.

- √ *Blended threats* typically will spread automatically by continuously scanning the Internet for Web servers with open or vulnerable

TCP/IP ports. They also plant Trojans and logic bombs, as well as change permissions and utilize internal network mapped drives to spread.

√ A virus with *stealth* characteristics will hide itself and send bogus responses back to an antiviral software package scan, in order to avoid detection.

√ A *polymorphic* virus is a virus that possesses the ability to change its own internal code and byte structure as it is being duplicated.

√ The *Backdoor.Subseven* virus and its known variants—Backdoor.Sub-Seven.1_7, Backdoor-G, Backdoor.Trojan, and Sub7—are most commonly distributed through e-mail attachments and instant messaging file and program transfers.

√ *NetBus* is a remote administration Trojan horse type program that is similar to Back Orifice and Backdoor.Subseven, which must first be executed on a system by a user in order to be installed.

√ *ILOVEYOU* is a self-propagating worm that is included as an e-mail attachment to an e-mail titled, "ILOVEYOU."

√ The *Melissa (W97M.Melissa.A)* virus is a macro virus that spreads very quickly when its payload is released or executed.

√ The *Back Orifice Trojan horse* is a program that is similar in nature to NetBus. It allows remote access to a computer system after a server application program has been executed on the remote or targeted computer system.

√ The *Chernobyl* virus, also named *W98.CIH* or just *CIH*, is an older *space filler* virus that targets earlier versions of operating systems such as Microsoft Windows 95 and Windows 98.

√ *W32.Nimda.A@mm* is a mass-mailing worm that targets the weaknesses of vulnerable, unpatched Microsoft IIS (Internet Information Server) Web servers.

√ *W32.Klez.A@mm* is a mass-mailing e-mail worm threat that exploits known weaknesses associated with Microsoft Outlook Express and Microsoft Outlook. (Remember what the "mm" means?)

√ If the Guest account is enabled in Microsoft Windows, it can be used to access shared resources without entering a password for authentication. This is a major security risk.

√ *Root* is the name of the administrative user account in UNIX and Linux.

CHAPTER SUMMARY

After completion of this chapter, you should have gained a good basic understanding of malicious code, specific computer viruses, antivirus solutions, and the very basics of operating systems security as they relate to the Security+ exam. At this point, you should hold yourself accountable for the following:

- You should know the differences and similarities between various type of malicious code such as viruses, worms, and Trojans.
- You should have the ability to identify security related terms such as Zoo, "in the wild," stealth, polymorphism, variant, payload, trigger, malware, spyware, and blended threat.
- You should be able to identify what type of malicious code or virus category specific viruses such as NetBus, Nimda, Brain, Backdoor.Trojan.dr(7), and BackOrifice are associated with. In other words, are they worms or Trojan horses, and so on?
- You should go to the recommended virus encyclopedia at *http://securityresponse.symantec.com/avcenter/vinfodb.html/* and become familiar with as many viruses as possible before taking the Security+ exam.
- For the exam, you should have an basic understanding of important antivirus software features such as enabling real-time protection, live updates, scheduled scans, filtering, and e-mail attachment scanning. For real life, you should have experience managing enterprise-wide antivirus solutions that include the earlier mentioned features as well as implementing centralized control features such as monitoring, quarantining, auto discovery, and server/client antivirus rollouts.
- You should be familiar with fundamental operating systems security practices and procedures inherent with Microsoft Windows and UNIX/Linux operating systems. If you have followed the recommendations included throughout this book, and used all the tools possibly available to you, you should now be well prepared for the real exam. However, actually passing the exam is completely in you hands. It would be in your best interest to take the exam as soon as you complete your study of this book.

Here's one final note: Technical certification is a tool that can help you prove your skills, acquire technical employment, and contribute to job advancement opportunities. However, there is absolutely no tool greater than hands-on experience!

REVIEW QUESTIONS

1. **This is considered a type of virus that can replicate itself. However, it does not attach to other programs. It can remain resident in memory and keep in contact with other segmented pieces of itself until triggered. What type of virus is it?**

   ○ A. Zoo.

   ○ B. Boot sector.

   ○ C. Macro.

   ○ D. Worm.

   ○ E. Trojan.

   ○ F. All of the above.

   ○ G. None of the above.

   Correct answer = D

   A worm is a type of virus that gets its name from its inherent ability to spread itself to other networked systems, remain resident in memory and keep in contact with other segmented pieces of itself until triggered by a certain event to the duplicate and spread itself. A Zoo virus threat is a threat that only exists in contained, controlled antivirus lab. System or boot infectors are viruses that are commonly known to attach themselves to and damage system files such as a hard drives Master Boot Record (MBR) or the boot sector on a floppy disk. A macro virus is virus that utilizes another applications, such as Microsoft Word or Excel. You were warned! It is very important to understand that while Worms and viruses duplicate themselves Trojans do not.

2. **This is used to describe viruses that are not contained within controlled environments. It describes viruses that commonly attack networks and systems that are used at home and in businesses on a regular daily basis. What is being described?**

   ○ A. In the Zoo.

   ○ B. In the Trojan.

○ C. In the wild.

○ D. In the Worm.

○ E. None of the above.

Correct answer = C

Viruses that are "in the wild" exist and spread on systems and in networks that are commonly used on a daily basis. Virus that are wild are also viruses that exist outside of contained and controlled environments, such as registered scientific antivirus research systems. All other choices are invalid.

3. **These types of viruses associate themselves with .com and .exe files. What are they?**

○ A. File infectors.

○ B. System or boot infectors.

○ C. Retroviruses.

○ D. Spyware.

○ E. Payload viruses.

○ F. None of the above.

Correct answer = A

File infectors are viruses that attach themselves to files that can typically be executed. They associate themselves with .exe and .com files. System or boot infectors are viruses that are commonly known to attach themselves to and damage system files, such as a hard drives, the Master Boot Record (MBR), or the boot sector on a floppy disk. A retrovirus is a virus that is designed to first attack antivirus software programs with the intent of passing through undetected. Spyware is a program or piece of software that resides hidden on a system that monitors and logs the systems or another systems activities. All other choices are invalid.

4. **This term represents unwanted or unwelcome code such as Trojans, viruses, and worms. What is this term?**

○ A. Macro viruses.

○ B. Freeware.

○ C. Shareware.

○ D. Malware.

○ E. All of the above.

Correct answer = D

Malware is shorthand for malicious code. Malware can be viruses, Trojans, or worms. Macro viruses utilize applications such as Microsoft Word or Excel, macro code, or programming language to be distributed. Freeware is software that is distributed by its copyright owner for free. The copyright owner usually retains the right to be the freeware's sole distributor. Shareware is software that is distributed freely by its owner. If you like it, you are supposed to pay for it.

5. **Which type of threat can cause mass damage to a network or system by combining the characteristics of several types of viruses and other malicious code?**

   ○ A. VWT threat.
   ○ B. Variant threat.
   ○ C. Blended threat.
   ○ D. Stealth threat.
   ○ E. None of the above.

Correct answer = C

Blended threats are viruses that combine the most lethal characteristics of viruses, worms, and Trojans in order to cause mass destruction. All other choices are invalid.

6. **This is something commonly left behind by disgruntled programmers, developers, or network administrators who have a grudge or a score to settle. What is it?**

   ○ A. Keys.
   ○ B. Virus threat rating.
   ○ C. Logic bomb.
   ○ D. Stealth bomb.
   ○ E. None of the above.

Correct answer = C

Disgruntled former employees with a grudge or a score to settle often leave behind logic Bombs. Although many folks are required to leave their office or other work related keys behind, this is not the appropriate answer to this question. A virus threat or risk rating is a calculated value that represents the possible level of severity or threat that an identified virus or piece of malicious code represents to a computer system. Answers D and E are invalid.

7. **Which of the following refers to mass-mailer worms that are attached to mail messages and can use a contact list or address book to widely be distributed?**

   ○ A. *.enc.*
   ○ B. *.dr.*
   ○ C. *@mm.*
   ○ D. *.mmm.*
   ○ E. *@enc.*
   ○ F. *@mdr.*
   ○ G. None of the above.

   Correct answer = C

   @mm refers to mass-mailer worms—viruses that attach themselves to malicious mail and are sent to contacts in an address or distribution list automatically. The . enc refers to a file that has been encrypted or encoded. The .dr refers to files that are considered to be dropper files. These are programs that drop a virus or worm onto a victim's computer system. All other choices are invalid.

8. **Which of the following is a mass-mailer worm whose payload is triggered on the 13th of most months?**

   ○ A. W32.Nimda.A@mm.
   ○ B. Chernobyl.
   ○ C. W97M.Melissa.A .
   ○ D. W32.Kriz.
   ○ E. W32.Klez.A@mm.
   ○ F. All of the above.
   ○ G. None of the above.

   Correct answer = E

   W32.Klez.A@mm is a mass-mailing e-mail worm threat that exploits known weaknesses associated with Microsoft Outlook Express and Microsoft Outlook. (Remember what the "mm" means?) The virus in known to release itself when the system date reached the 13th in January, March, May, July, September, and November.

W32.Nimda.A@mm proliferates through e-mail attachments and uses the Unicode Web Traversal exploit. It is not specifically triggered on the 13th of most months. Chernobyl or W98.CIH or just CIH, is an older space filler virus that targets earlier versions of operating systems such as Microsoft Windows 95 and Windows 98. It is triggered on April 26th. The Melissa (W97M.Melissa.A) virus is a macro virus that spreads very quickly when its payload is released or executed. The W32.Kriz virus is a virus similar to the Chernobyl virus, which typically resides in computer systems memory. Its payload is triggered on December 25th of most years.

9. **Which of the following are considered to be Trojan horse remote administrative threats that exist in the wild? (Choose three)**
   - ☐ A. NetBus.
   - ☐ B. Nimda.
   - ☐ C. Backdoor.Subseven.
   - ☐ D. ILOVEYOU.
   - ☐ E. Back Orifice.
   - ☐ F. None of the above.

   Correct answers = A, C, and E

   The Backdoor.Subseven, Back Orifice, and NetBus are considered to be Trojan horse remote administrative threats that exist in the wild. All other choices are invalid.

10. **Where are unencrypted passwords stored in UNIX?**
    - ○ A. */etc/shadow* file.
    - ○ B. *C:\winnt\system32\drivers\etc.*
    - ○ C. */etc/passwd* file.
    - ○ D. */dev.*
    - ○ E. */tmp.*
    - ○ G. None of the above.

    Correct answer = C

    By default, unencrypted passwords are stored in the /etc/passwd file using UNIX or Linux. Encrypted passwords are stored in the /etc/shadow file. C:\winnt\system32\drivers\etc is a location used to hold files in Windows operating Systems. The /dev area contains files that point to the physical devices that are attached to the system using UNIX or Linux. The /tmp directory stores temporary files and is referred to as the scratch area.

11. Which command is used for backup purposes in UNIX and Linux?

    ○ A. *Backup.*

    ○ B. *Archive.*

    ○ C. *tar.*

    ○ D. *kill.*

    ○ E. *rlogin.*

    ○ G. None of the above.

Correct answer = C

In UNIX or Linux, tar is similar to the DOS copy command; it is used for archival and backup purposes. The kill command is used to terminate application processes. The rlogin command is used to initiate remote access sessions. All other choices are invalid.

12. What is the administrator account called in Windows?

    ○ A. A security weakness.

    ○ B. Admin.

    ○ C. Supervisor.

    ○ D. *root.*

    ○ E. Administrator.

    ○ G. None of the above.

Correct answer = E

In Windows, the default account with administrative privileges is called Administrator. Older Novell operating system versions use Supervisor as the administrative account. In UNIX or Linux, root is the administrator equivalent.

REFERENCES

http://www.wildlist.org is provided by the Wild List Organization International. It lists the "in the wild" viruses that exist.

http://securityresponse.symantec.com/avcenter/vinfodb.html/ is offered by Symantec. It has an extensive virus encyclopedia that can keep you up-to-date on the most resent as well as past virus threats.

http://securityresponse.symantec.com is the Symantec Security Response Web site. It lists the latest virus definition files, virus threats, and available virus removal tools.

THE SECURITY+ EXAM GUIDE TESTTAKER'S CUMULATIVE PRACTICE EXAM

The questions in this practice exam are based on topics that have been presented to you in Chapters 2 through 9 of this book. The answers to the following questions as well as the chapter headers they are taken from are provided at the end of the exam.

ON THE CD If you can answer all of the questions in this practice exam and the practice exams included with the accompanying CD-ROM correctly, as well as understand the theory behind each of the topics mentioned, there is a good possibility that you may pass the Comp-TIA Security+ examination.

NOTE This exam is more difficult than the real Security+ exam. So suit up, put your foot soldier's armor on, pass this test, and then go take the real one. Good luck!

1. **Which of the following suggestions should not be taken when creating a strong password?**

   ○ A. A strong password should be a minimum of seven characters in length.

   ○ B. The password should contain a combination of upper and lowercase letters.

   ○ C. There should be at least one number contained in the password.

   ○ D. The password should contain at least one of the following characters: !@#$%^&*.

   ○ E. A password should be a word from a dictionary, person's name, family name, phone number, birth date, or favorite phrase.

   ○ F. None of the above.

2. **Which of the following best describes two-factor authentication?**

   ○ A. Combining a physical trait or a possession (magnetic card, key, and so on) with a password or PIN.

   ○ B. Two User IDs.

   ○ C. Two security guards.

   ○ D. Two passwords.

   ○ E. SSO.

   ○ F. All of the above.

3. **Which access control security model supports the following statement?**

   "If user B has a lower security level than User A, User B should not be able to write over User A's information."

   ○ A. Clark-Wilson/

   ○ B. Biba.

   ○ C. Bell-LaPadula.

   ○ D. Toback-Crayton.

   ○ E. Non-interference.

   ○ F. All of the above.

4. **Which of the following can allow users access to enterprise-wide resources with one sign-on?**

   ○ A. Open source.

   ○ B. Netlogon Service.

   ○ C. SSO.

   ○ D. SESAME.

   ○ E. Relativ6 authentication.

   ○ F. All of the above.

5. **Which of the following is currently the most secure biometric device available?**

   ○ A. Fingerprint scanner.

   ○ B. Retina scanner.

   ○ C. Encrypted-key scanner.

   ○ D. Brain scanner.

   ○ E. Token scanner.

   ○ F. None of the above.

6. Which of the following is not a valid file system access right?
   - A. Read.
   - B. Write.
   - C. Delete.
   - D. Execute.
   - E. Undelete.
   - F. All are valid.

7. Which access control type is based on the roles users play within an organization?
   - A. Task-based.
   - B. Mandatory.
   - C. DACL.
   - D. Lattice-based.
   - E. Role-based.
   - F. RADIUS.
   - G. Rule-based.
   - H. None of the above.

8. Which of the following is a protocol used to transfer authentication and authorization data between a dial-in client and a server that supports PAP, CHAP, PPP, and UNIX login authentication methods?
   - A. TBAC.
   - B. LBAC.
   - C. DACL.
   - D. MAC.
   - E. RADIUS.
   - F. None of the above.

9. Developed by Cisco©, which remote authentication protocol is typically implemented on UNIX servers?
   - A. TBAC.
   - B. LBAC.
   - C. TACACS+.
   - D. RADIUS.
   - E. None of the above.

10. **Which of the following services and protocols should you consider removing from a system if they are not needed?**

    ○ A. Computer Browser service.

    ○ B. IIS Admin service.

    ○ C. FTP Server service.

    ○ D. Spooler service.

    ○ E. Netlogon service.

    ○ F. DHCP.

    ○ G. All of the above.

11. **When someone pretends to be someone else in order to gain information or unauthorized access, what are they practicing?**

    ○ A. Trojan-like behavior.

    ○ B. Spoofing.

    ○ C. Public persuasion.

    ○ D. Good people skills.

    ○ E. Social engineering.

    ○ F. None of the above.

12. **If you are using UNIX, it would be considered good practice to audit which of the following items?**

    ○ A. *.rhosts.*

    ○ B. */etc/password.*

    ○ C. Use of Setgid.

    ○ D. Bin files.

    ○ E. All of the above.

13. **By default, where are passwords stored in UNIX?**

    ○ A. *.rhosts.*

    ○ B. */etc/password.*

    ○ C. */etc/bin.*

    ○ D. */rhosts/bin/password.*

    ○ E. *.hosts.*

    ○ F. None of the above.

14. **Which of the following guidelines should you follow when using RAS?**
    - ○ A. Users should use different passwords for RAS dial in than they do for domain authentication.
    - ○ B. The time frame users can log in should be limited to business need.
    - ○ C. Callback security should be used when at all possible.
    - ○ D. RAS server and your remote clients should always have the latest operating system service packs.
    - ○ E. All of the above.

15. **What does 802.1X technology use that enables it to work with wireless, Ethernet, and Token Ring networks?**
    - ○ A. Carrier Sense Multiple Access/Collision Detection (CSMA/CD).
    - ○ B. Lightweight Directory Access Protocol (LDAP).
    - ○ C. Extensible Authentication Protocol (EAP).
    - ○ D. Synchronous Data Link Control (SDLC).
    - ○ E. None of the above.

16. **What is the main function of S/MIME?**
    - ○ A. It provides authenticity and privacy for e-mail messages in MIME format.
    - ○ B. It is used to scan IMAP and POP3 mail services.
    - ○ C. It allows RSA algorithms to encrypt a session key with a receiver's public key.
    - ○ D. It allows antivirus programs to scan e-mail attachments.
    - ○ E. None of the above.

17. **Which of the following are commonly used security protocols that provide transport security through Internet browsers provided by Netscape and Microsoft?**
    - ○ A. HTTPS and IPX/SPX.
    - ○ B. SSL and TLS.
    - ○ C. TCP/IP and SMTP.
    - ○ D. FTP and S/FTP.
    - ○ E. None of the above.

18. **Which technique is often used to quarantine applets that appear suspicious or malicious?**

   ○ A. Script pots.
   ○ B. Sandboxing.
   ○ C. Applet sniffing.
   ○ D. Data mining.
   ○ E. None of the above.

19. **Which is the most secure type of FTP?**

   ○ A. FTSP.
   ○ B. TFTP/S.
   ○ C. S/FTP.
   ○ D. FTP/S.
   ○ E. None of the above.

20. **Certified engineers that can properly recommend and assist you with integration plans as well as keep you in line with federal, state, and local regulations regarding wireless networks will typically perform what?**

   ○ A. A corporate security analysis.
   ○ B. An antivirus scan.
   ○ C. A host-based intrusion analysis.
   ○ D. A professional site survey.
   ○ E. None of the above.

21. **This layer establishes, holds, and controls connections between two applications. It provides checkpoint and synchronization service between two communication connections. Security between two connections is also handled at this layer. Which OSI layer is being described?**

   ○ A. Application.
   ○ B. Presentation.
   ○ C. Session.
   ○ D. Transport.
   ○ E. Network.
   ○ F. Data Link.
   ○ E. Physical.

22. **Which statement is not true concerning FDDI rings?**

    ○ A. Repeaters are used to boost the signal on an FDDI ring.

    ○ B. An FDDI ring in a LAN environment has a distance capability of up to 200km (124 miles).

    ○ C. FDDI rings often serve as network backbones for WANs.

    ○ D. FDDI uses token passing technology.

    ○ E. None of the above statements is true.

23. **When part of an internal network has been made accessible to outside sources, that part of the internal network is referred to as?**

    ○ A. An Internet.

    ○ B. An intranet.

    ○ C. An extranet.

    ○ D. A VLAN.

    ○ E. A security hole.

24. **Of the following, which is the process of establishing a known footprint or baseline of a system's usage of such things as CPU (Central Processing Unit) utilization, disk utilization, use of user rights, user logins, file and folder access over time, and analyzing the system for any deviation from the system's baseline or "normal" behavior?**

    ○ A. Network-Based IDS (NIDS).

    ○ B. Host-Based IDS (HIDS).

    ○ C. Signature Intrusion Analysis.

    ○ D. Statistical Intrusion Analysis.

    ○ E. None of the above.

25. **A solid incident response procedure should include which of the following items?**

    ○ A. Written procedures.

    ○ B. Actual steps that will be implemented to correct, repair, or restore whatever has been damaged.

    ○ C. Who will be notified, how they will be notified, and when they will be notified

    ○ D. A written sign off that the plan was tested.

    ○ E. All of the above.

26. The advantage of this type of router is its ability to analyze and drop or pass packets quickly. The major disadvantage of this technology is that packets are not typically analyzed beyond source and destination addresses. What type of firewall or router is being described?

    ○ A. A Stateful inspection firewall.
    ○ B. A packet-filtering router.
    ○ C. A multi-homed gateway.
    ○ D. An application proxy.
    ○ E. None of the above.

27. Which of the following best describes DMZ?

    ○ A. It is a dedicated switching technology that transmits data in fixed length 53-byte units called cells.
    ○ B. It is a neutral area between an internal network and the Internet that typically contains one host system or a small network of systems.
    ○ C. It is used to hide internal network IP addresses from external network sources.
    ○ D. It is much faster than X.25 and can take advantage of T1 (1.544 Mbps) and T3 (Mbps) speeds.

28. Which of the following are not physical controls that are not commonly implemented in order to place a "barrier" or form of protection between unauthorized personal and sensitive locations or data?

    ○ A. Turnstiles.
    ○ B. Mantraps.
    ○ C. Guards.
    ○ D. Dogs.
    ○ E. HVAC systems.
    ○ F. All are physical controls.

29. Which of the following are important considerations when determining a new site location for your company or business?

    ○ A. A site located in an area with a low crime rate.
    ○ B. A geographically stable site. For example, no fault lines, low flood area, no trash dumps.

○ C. A site with multiple access paths into and out of the site location.

○ D. A site located away from airline, railway, and major construction paths.

○ E. All of the above.

○ F. None of the above.

30. **Which of the following types of extinguishers are the most common and are considered effective at extinguishing chemical, electrical, and normal wood burning fires?**

○ A. Type A.

○ B. Type ABC.

○ C. Type BC.

○ D. Halon.

○ E. Carbon Dioxide (CO_2).

○ F. None of the above.

31. **Which backup strategy is the most efficient for backups and that includes a daily, weekly, and monthly backup schedule?**

○ A. Incremental.

○ B. Tar.

○ C. GFS (Grandfather-Father-Son).

○ D. Full.

○ E. Differential.

○ F. Copy.

○ G. None of the above.

32. **This provides excellent fault tolerance and good performance. It stores parity information across all disks in the disk array and provides concurrent disk reads and writes. It is the most popular RAID implementation. Which RAID level is it?**

○ A. 0.

○ B. 1.

○ C. 3.

○ D. 5.

○ E. 32.

○ F. 64.

33. **The grouping together of independent servers into one large logical system is called what?**
    - ○ A. Collective engineering.
    - ○ B. Server collaboration.
    - ○ C. Server clustering.
    - ○ D. Logical conglomeration.
    - ○ E. RAID Web farming.
    - ○ F. None of the above.

34. **In order to store passwords securely, some systems only maintain a copy of the password's _____.**
    - ○ A. Hop.
    - ○ B. Cluster.
    - ○ C. Key length.
    - ○ D. Hash.
    - ○ E. Bit length.
    - ○ F. None of the above.

35. **One advantage of Digital Signatures (DSes) is that they can bind a person to an agreement. Which of the following describes what this aspect of a DS provides?**
    - ○ A. Non-negotiation.
    - ○ B. Non-repudiation.
    - ○ C. Non-disclosure.
    - ○ D. *Nolo contendere.*
    - ○ E. Non-stipulation.
    - ○ F. None of the above.

36. **Which of the following best describes the method used by an algorithm to produce subkeys?**
    - ○ A. Key-mating.
    - ○ B. Key-division.
    - ○ C. Key-branching.
    - ○ D. Key-splitting.
    - ○ E. Key-scheduling.
    - ○ F. None of the above.

37. The cryptographic strength of the Diffie-Hellman algorithm is based on which mathematical problem?

- ○ A. Prime number.
- ○ B. Discrete logarithm.
- ○ C. Factoring.
- ○ D. Long division.
- ○ E. Logistic discretion.
- ○ F. None of the above.

38. Which of the following is a disadvantage of incorporating DNSSEC with a DNS server?

- ○ A. Enables masquerading.
- ○ B. Domain names cannot be resolved to IP addresses.
- ○ C. Database size is increased.
- ○ D. Uses smaller key sizes resulting in weaker encryption.
- ○ E. Adding DNS protocol extensions can reduce database integrity.
- ○ F. All of the above.

39. Which standard does a typical digital certificate conform to?

- ○ A. X.509.
- ○ B. ANSI.
- ○ C. NIST.
- ○ D. DES.
- ○ E. RFC 822.
- ○ F. RFC 733.

40. If a company maintains copies of their employee's crypto-keys, which type of key should they revoke first if the employee were to quit?

- ○ A. Encryption key.
- ○ B. Subkey.
- ○ C. Shared key.
- ○ D. Signing key.
- ○ E. Access-control key.
- ○ F. Escrow key.

41. **Which component of an expert system acts as an intermediary and searches a knowledge base for patterns that are relevant to the user's query?**

    ○ A. AI.
    ○ B. Neuron.
    ○ C. Parallel processor.
    ○ D. Rule-based program.
    ○ E. Inference engine.
    ○ F. Paradigm.

42. **Which type of KBS attempts to emulate biological systems like the human brain?**

    ○ A. Pattern matching.
    ○ B. Rule of thumb.
    ○ C. Waterfall.
    ○ D. Neural network.
    ○ E. Expert system.
    ○ F. Fountain.

43. **In the SDLC, which of the following describes user participation during beta testing?**

    ○ A. Expert testing.
    ○ B. Inference testing.
    ○ C. Acceptance testing.
    ○ D. Disposition testing.
    ○ E. Novice testing.
    ○ F. None of the above.

44. **In a database, what is the function of a primary key?**

    ○ A. Uniquely identifies records in a table.
    ○ B. Encrypts the database.
    ○ C. Connects cells.
    ○ D. Connects tuples.
    ○ E. Digs into a data mine.
    ○ F. None of the above.

45. **Web spiders that crawl around the Internet looking for data are known as which type of application?**
    - ○ A. Applets.
    - ○ B. Viruses.
    - ○ C. Agents.
    - ○ D. ActiveX.
    - ○ E. Objects.
    - ○ F. Neural networks.

46. **One method of hardening a mail server is to implement secure POP3 access. Which protocol is used to encrypt POP3 logons?**
    - ○ A. SSH.
    - ○ B. SSL.
    - ○ C. DSS.
    - ○ D. SMTP.
    - ○ E. IMAP.
    - ○ F. None of the above.

47. **The use of polyinstantiation without the proper database-wide updating can cause a loss in data _____.**
    - ○ A. Confidentiality.
    - ○ B. Encryption.
    - ○ C. Repudiation.
    - ○ D. Translation.
    - ○ E. Integrity.
    - ○ F. All of the above.

48. **In a rule-based expert system, once a rule has been matched to a user's query, which parameter defines the action to be taken?**
    - ○ A. If.
    - ○ B. Then.
    - ○ C. Act.
    - ○ D. And.
    - ○ E. Or.
    - ○ F. Def.

49. **Which of the following computer laws prohibits eavesdropping by way of wire or oral communications?**

    ○ A. 1968 Abycus Act .
    ○ B. 1974 Federal Privacy Act.
    ○ C. 1986 Electronic Communications Privacy Act.
    ○ D. 1987 Computer Security Act.
    ○ E. None of the above.

50. **Which of the following is the most commonly committed computer crime?**

    ○ A. Data diddling.
    ○ B. Business and financial attacks.
    ○ C. Embezzlement.
    ○ D. Fraud.
    ○ E. Salami attacks.
    ○ F. Pepperoni attacks.
    ○ G. Grudge attacks.

51. **In order for a company to have or maintain an advantage over its competitor, what must not be divulged?**

    ○ A. Trademarks.
    ○ B. Business and financial policies.
    ○ C. Trade secrets.
    ○ D. Patents.
    ○ E. Copyrights.
    ○ F. None of the above.

52. **This type of attack is not usually launched with the intent of producing a financial gain or causing harm. Younger enthusiasts such as college students and script kiddies most often carry this type of attack out. What type of attack is being described?**

    ○ A. Grudge attack.
    ○ B. Postgraduate attack.
    ○ C. Hearsay attack.
    ○ D. Adolescent script attack.
    ○ E. Fun attack.
    ○ F. Data diddling.

53. **When does the chain of evidence custody begin?**

   ○ A. After the proper authorities have been contacted.

   ○ B. When the first responsible person alerts management.

   ○ C. When the first piece of evidence is collected.

   ○ D. When the intruder is identified.

   ○ E. When the chain of evidence custody report is first documented.

   ○ F. None of the above.

54. **Which of the following is located at Purdue University and is one of the largest computer security research groups in the world?**

   ○ A. AIB.

   ○ B. TPUCSRG.

   ○ C. Gramm, Leach, and Bliley.

   ○ D. IAB.

   ○ E. COAST.

   ○ F. None of the above.

55. **Which of the following are good security practices when securing Microsoft Windows?**

   ○ A. Disable FTP services.

   ○ B. Enable auditing.

   ○ C. Rename the Administrator account.

   ○ D. Set strong passwords on all accounts, and mix characters and numbers.

   ○ E. Disable the Guest account (and other unnecessary accounts).

   ○ F. All of the above.

56. **Which of the following is usually associated with disgruntled employee syndrome, is also referred to as *slag code*, and is most often activated when a certain time is reached or a certain event occurs?**

   ○ A. Trojan horse.

   ○ B. Worm.

   ○ C. Logic bomb.

   ○ D. Stealth horse.

   ○ E. Polymorphic worm.

   ○ F. None of the above.

57. **Which of the following is not a remote administration, Trojan horse-style application?**

   ○ A. W32.Nimda.A@mm.
   ○ B. NetBus.
   ○ C. Backdoor.NetBus.
   ○ D. BackOrifice.
   ○ E. BO Trojan.
   ○ F. None of the above.

58. **This was considered to be one of the first stealth viruses. It would first hide itself to avoid detection by virus scanners. It would then infect a hard drives boot sector. What was this virus called?**

   ○ A. Nimda.
   ○ B. admiN.
   ○ C. Backdoor7.
   ○ D. The Brain.
   ○ E. BO Trojan.
   ○ F. W32.Nimda.A@mm.
   ○ G. None of the above.

59. **Which of the following topologies is considered the most redundant?**

   ○ A. Full-bus.
   ○ B. Partial-star.
   ○ C. Full-mesh.
   ○ D. Partial-mesh.
   ○ E. Full-star.
   ○ F. All of the above are equally redundant.

60. **Digital signatures can provide proof that a transaction or contract has occurred so that it is not denied at a later time. What is being described here?**

   ○ A. Vulnerability.
   ○ B. Public/private key pairs.
   ○ C. Weak keys.
   ○ D. Non-repudiation.
   ○ E. Cryptography.

61. **What is used to limit a subject's access to an object?**

    ○ A. Access right.

    ○ B. Buffer.

    ○ C. DOS.

    ○ D. Hash.

    ○ E. SATAN.

62. **Which of the following is a computer program (virus) designed to duplicate itself again and again until it exhausts system resources?**

    ○ A. Cipher.

    ○ B. Worm.

    ○ C. Trojan horse.

    ○ D. Stealth.

    ○ E. All of the above.

    ○ F. None of the above.

63. **These can be used to monitor network traffic and weaknesses and can also be used to steal passwords, user IDs, or credit card information. What are they?**

    ○ A. Firewalls.

    ○ B. Auditors.

    ○ C. Loggers.

    ○ D. Sniffers.

    ○ E. Enraptures.

    ○ F. All of the above.

64. **Hijackers and attackers often create or intercept these and use them to manipulate files on users computers. What are they?**

    ○ A. User profiles and system policies.

    ○ B. Fingerprint and retina patterns.

    ○ C. Java scripts and applets.

    ○ D. Tokens and SAMS.

    ○ E. Policies and permissions.

    ○ F. DTEs and DCEs.

    ○ G. All of the above.

65. **Which of the following represent benefits of implementing Single Sign-Ons?**

    ○ A. Users only have to sign on one time to access enterprise-wide resources.

    ○ B. The risks involved with individual's users having to remember multiple user IDs and passwords are eliminated or reduced.

    ○ C. Administrative overhead is reduced.

    ○ D. None of the above.

    ○ E. All of the above.

66. **(BONUS QUESTION!) It is likely that the Security+ exam will expect you to have knowledge of which of the following viruses and the major virus category types they are associated with?**

    ○ A. NetBus.

    ○ B. Nimda.

    ○ C. Brain.

    ○ D. Backdoor7.

    ○ E. BackOrifice.

    ○ F. All of the above.

CUMULATIVE PRACTICE EXAM ANSWERS

Answer Key	*Question Taken From*
1. E	Chapter 2, "Authentication"
2. A	Chapter 2, "Authentication"
3. B	Chapter 2, "Access Control Systems Fundamentals"
4. C	Chapter 2, "Authentication"
5. B	Chapter 2, "Authentication"
6. E	Chapter 2, "Authorization"
7. E	Chapter 2, "Access Control Techniques"
8. E	Chapter 2, "Access Control Techniques"
9. C	Chapter 2 "Access Control Techniques"
10. E	Chapter 2, "Nonessential Services and Protocols"
11. E	Chapter 2, "Attack Methods"
12. E	Chapter 2, "Auditing"
13. B	Chapter 9, "UNIX/Linux Security"
14. E	Chapter 3, "RAS (Remote Access Service)"
15. C	Chapter 3, "802.1X"
16. A	Chapter 3, "E-mail Security"
17. B	Chapter 3, "Web Security"
18. B	Chapter 3, "Web Security"
19. C	Chapter 3, "File Transfer"
20. D	Chapter 3, "Wireless Security"
21. C	Chapter 4, "Open Systems Interconnection (OSI) Model and Layers"
22. A	Chapter 4, "Network Topology"
23. C	Chapter 4, "Network Topology"
24. D	Chapter 4, "Intrusion Detection Systems (IDSes)"
25. E	Chapter 4, "Intrusion Detection Systems (IDSes)"
26. B	Chapter 4, "Network Countermeasures"
27. B	Chapter 4, "Network Countermeasures"
28. E	Chapter 6, "Physical Security"
29. E	Chapter 6, "Physical Security"
30. B	Chapter 6, "Physical Security"
31. C	Chapter 6, "Disaster Recovery"

32. D	Chapter 6, "Business Continuity"
33. C	Chapter 6, "Business Continuity"
34. D	Chapter 5, "Algorithms"
35. B	Chapter 5, "Concepts of Using Cryptography"
36. E	Chapter 5, "Algorithms"
37. B	Chapter 5, "Algorithms"
38. C	Chapter 5, "Standards and Protocols"
39. A	Chapter 5, "Public Key Infrastructure (PKI)"
40. D	Chapter 5, "Key Management/Certificate Lifecycle"
41. E	Chapter 7, "Knowledge-based Systems"
42. D	Chapter 7, "Knowledge-based Systems"
43. C	Chapter 7, "System Development Life Cycle (SDLC)"
44. A	Chapter 7, "Databases"
45. C	Chapter 7, "Applications"
46. B	Chapter 7, "Applications"
47. E	Chapter 7, "Databases"
48. B	Chapter 7, "Knowledge-based Systems"
49. C	Chapter 8, "Law"
50. A	Chapter 8, "Computer Crime"
51. C	Chapter 8, "Law"
52. E	Chapter 8, "Computer Crime"
53. C	Chapter 8, "Investigation"
54. E	Chapter 8, "Investigation"
55. F	Chapter 9, "Operating Systems Security"
56. C	Chapter 9, "Computer Virus Types"
57. A	Chapter 9, "Computer Virus Types"
58. D	Chapter 9, "Computer Virus Types"
59. C	Chapter 4, "Network Topology"
60. D	Chapter 2, "Basic Terminology"
61. A	Chapter 2, "Basic Terminology"
62. B	Chapter 2 "Basic Terminology" and Chapter 9, "Computer Virus Types"
63. D	Chapter 2, "Test Tips"
64. C	Chapter 3, "Test Tips"
65. E	Chapter 6, "Privilege Management"
66. F	Chapter 9, "Computer Virus Types"

ABOUT THE CD-ROM

The CD-ROM included with this book contains Security+ practice exams that will prepare you well for the CompTIA Security+ certification examination.

It is highly recommended that you keep taking these exams until you pass each one every time. This will ensure your best chances for passing the real CompTIA Security+ examination.

REQUIREMENTS

You will need Internet Explorer 5.0 or greater with JavaScript enabled. To download IE 6.0 go to
http://www.microsoft.com/windows/ie/downloads/ie6/default.asp.

The exam will run under Windows 95 (all versions), Windows 98 (all versons), Windows NT 4.0 workstation, Windows NT 4.0 Server, Windows 2000 Professional, Windows 2000 Server, Windows 2000 Advanced Server, Windows Me, Windows XP Home, and Windows XP Professional.

INSTALLATION

No installation is required. Simply insert the CD-ROM, navigate to your CD-ROM drive letter, and double-click **Security+ Exams.hta** to start the program. You can copy the program to your hard drive by copying the program file and Include folder to a local drive.

GENERAL OPERATION

There are four Security+ TestTaker's exams on the CD-ROM included with this book. Each exam contains 30 questions and must each be completed within 30 minutes. You will receive a score of pass or fail at the end of each exam. You must answer 25 of the 30 questions correctly in order to pass each exam. If you do not answer all 30 of the required questions contained in each of the individual exams within 30 minutes, you will fail the exam. At the end of each exam, a review option is available to check your answers.

KEYBOARD SHORTCUTS

Two keyboard shortcuts are available once a practice exam has been started.

To navigate to the next question after starting a practice exam use Alt + N for Next. To go back to the previous question use Alt + B for Back.

GLOSSARY

access control   To limit the use of an object or subject that has been authorized. Methods placed to limit the access to resources from unauthorized users or programs.

API (Application Program Interface)   A set of uniform routines or rules that allow programmers and developers to write applications that can be used to interact with various operating system platforms. APIs define system calls for service.

ARP (Address Resolution Protocol)   A TCP/IP protocol used to determine the hardware MAC address for a network interface card.

asymmetric algorithm   An encryption process that uses a pair of keys to securely encrypt and decrypt messages so that they arrive to an intended receiver safely.

authentication   A method used to verify the identity of a user or subject to a system. Authentication is typically a prerequisite for access to a system resource.

authorization   Permission granted to a subject to access or utilize a particular object such as a file or folder.

back door   A hidden entry point to a program or system usually created by the application or system manufacturer. Back doors provide access to a system that typically exploited by unauthorized attackers.

bastion host   A system that has been protected or hardened in preparation for an expected attack. A bastion host is used to protect an internal network from external attacks.

BIOS (Basic Input/Output System) The BIOS is software built into a ROM BIOS or flash BIOS chip that is used to control hardware devices such as hard drives, keyboards, monitors, and other low-level devices before a computer system boots into an operating system.

BNC (Bayonet Nut Connector, Bayonet Neil-Concelman, or British Naval Connector) A connector used to connect a computer to a coaxial cable in a 10base2 Ethernet network.

BPS (bits per second) A standard measurement of the speed at which data is transmitted. For example, a 56K modem has the ability to transmit at a rate of 56,000 bits per second.

brute force A program that tries to all possible characters and phrase combinations in order to gain a password or PIN that can be used to illegally authenticate.

CD (compact disk) A round metallic disk that stores information such as text, video, and audio in digital format.

CD-R (compact disk-recordable) A type of compact disk that can be written or recorded to once but read many times.

CERT The CERT (Computer Emergency Response Team) was established at Carnegie-Mellon University. Its goal is to provide useful information that can assist with current and future security problems and threats.

CHAP (Challenge Handshake Authentication Protocol) Authentication method where a server system sends a client system a randomly selected value and ID. CHAP uses a one-way hash value that is typically created using MD5. CHAP is mush more secure that PAP.

clustering The grouping of individual systems together allowing them to act as one large system. Clustering allows multiple servers the ability to access single disk arrays that contains applications and services.

confidentiality Making sure that only authorized individuals or systems have access to data.

cracker One who breaks into a secured system with malicious intent. Crackers most commonly use brute force and dictionary attack methods as tools to figure out passwords.

CSMA/CD (Carrier Sense Multiple Access/Collision Detection) A contention-based protocol used to detect collisions of packets in

Ethernet networks. If a collision occurs the information is retransmitted.

decrypt To convert enciphered text to into plain text.

DES (Data Encryption Standard) A 56-bit symmetric-key encryption method.

DHCP (Dynamic Host Configuration Protocol) A protocol used to assign IP addresses dynamically to computer systems in a TCP/IP network. DCHP eases administrative overhead by reducing the need to assign individual static IP addresses.

DHTML (Dynamic Hypertext Markup Language) A new form of HTML programming code that allows developers to create more interactive/responsive Web pages for users.

digital signature An electronic version of a signature used to authenticate and identify the sender of information. Primarily used for identification purposes and the prevention of forgery.

DNS (Domain Name System) An Internet service that translates fully qualified domain names to computer IP addresses.

DNS spoofing Pretending to be a valid DNS server by stealing its domain name or compromising the target DNS server name cache.

DoS (Denial of Service) The loss of resources that are normally available. Malicious attack programs typically aim to deny access to these resources for normal users.

DOS (Disk Operating System) A 16-bit operating system developed by Microsoft that does not support true multitasking capabilities.

DSL (Digital Subscriber Line) A popular high-speed technology that uses phone lines for Internet connectivity. The two most widely used forms of DSL are ADSL (asynchronous) and SDSL (synchronous).

due care Acting responsibly and in good faith. Acting in the best interest of one's company or enterprise.

encryption The process of changing or concealing data or programs so that they cannot be viewed in plain text. This is usually accomplished through the use of an algorithmic program.

firewall Software or hardware, or a combination of both, designed to prevent access to internal networks and resources from outside sources. A firewall is usually installed on a server that acts as gateway or router. A firewall looks at data packets and screens them for validity.

FTP (File Transfer Protocol)   A transfer protocol primarily used on the Internet to transfer files from one location to another.

fault tolerance   The ability of a program or system to remain functional in the event of a hardware or software failure. There are various levels of fault tolerance that offer different levels of protection. The most common are RAID (Redundant Array of Independent disks) levels 1, 3, and 5.

GB (gigabyte)   A measurement of computer system data storage data storage space. One GB is equal to 1,024 megabytes or approximately 1 million kilobytes.

hacker   An expert computer-programming enthusiast who has the knowledge and capabilities to gain unauthorized access to secured computer systems and programs.

hash   A value that is that is generated from a string of characters or text. It is very unlikely that a duplicate hash value will ever be produced from various strings of text. Hashing is the changing or transforming of a set of characters into a shorter set or value of numbers. A hashing algorithm known as a hash function is used to disorganize values to make them more difficult to figure out. Hashing is often used with the encryption and decryption of digital signatures.

HTML (HyperText Markup Language)   A programming language that is used to create pages or hypertext documents on the World Wide Web. HTML is a scripting language that uses tags to define the way Web pages are displayed.

HTTP (HyperText Transfer Protocol)   A fast Internet application protocol used for transferring data.

IEEE (Institute of Electronic and Electrical Engineers)   The world's leading international standards organization that's primary purpose is the development of Information Technology (IT) standards and the welfare of its members.

IETF (International Engineering Task Force)   An important Internet body or society that provides standards and Internet protocols. These standards are known as RFCs (Requests For Comments). RFCs can become documented procedures or actual standards.

intrusion detection   A security management system used to identify security weaknesses and breaches that are internal or external to a network. This type of system should gather, analyze and assess security information that relates to vulnerabilities that are dangerous to normal operations.

IP (Internet Protocol)   A TCP/IP protocol used primarily to allow computers to be connected in a local area network or to the Internet.

IPX/SPX (Internetwork Packet Exchange/Sequence Packet Exchange)   A Novell networking protocol suite used primarily with Novell Netware.

ISDN (Integrated Services Digital Network)   A digital communications standard that allows data and voice to be used on the same phone line connection. ISDN provides support for up to 128kbps transfer rates and is intended to replace traditional analog technology.

ISP (Internet service provider)   A company whose primary business is to provide access to the Internet for other companies and individuals.

Kbps (kilobits per second)   A measurement of data transfer rate. One kbps is equivalent to 1,000 bits per second.

KB (kilobyte)   1,024 bytes.

LAN (local area network)   A network of computers that are typically connected together in a central location such a building. In a LAN, computers are connected together by wires or other media and share common resources such as printers, files, and modems.

logging   The process or actions implemented to store information that is obtained from networked workstations, servers, or firewalls.

MAPI (Messaging Application Programming Interface)   A Microsoft application-programming interface that provides the ability to send e-mail and attachments from within programs such as Word, Excel, PowerPoint, and Access.

MB (megabyte)   1,024 kilobytes or 1,048,576 bytes.

modem (modulator demodulator)   A communications device used to convert signals so they can be transmitted over conventional telephone lines. A modem converts incoming analog signals to digital format and outgoing digital signals to analog format.

NIC (network interface card)   An electronic circuit board that attaches a computer to a network. A NIC is installed inside a computer connects to a wire that typically leads to a networked hub, router, or bridge.

NTFS (NT File System)   A Windows NT hard drive file system that offers file and object level security features, file compression, encryption, and Long file Name support. A new version of the NTFS file system called NTFS5 is offered with the Windows 2000 operating system.

OLE (object linking and embedding)   A specification created by Microsoft that allows objects that are created in one program or application to be embedded or linked to another applications. With OLE, if a change is made to an application, the change is also made to the second application.

OSI (Open Systems Interconnection)   The OSI reference model is a networking model developed to provide network designers and developers with a model that describes how network communication takes place.

OTP (One Time Password)   A password that is used only once. This password cannot be stolen or used multiple times. It is considered secure and assists against password stealing programs.

PDA (personal data assistant)   A small handheld mobile computing device that provides functions similar to a desktop or laptop computer. Most PDAs today use a Pen/Stylus in place of a keyboard to input data.

policy   A company or organizations set of rules that provide guidelines that describe proper use of company assets, security rules, and company procedures.

private key   One part of a two-part cryptographic key, used to exchange private or secret messages between privy parties by encrypting or decrypting such message. Can be used to produce a Digital Signature.

proxy   Allows revisited pages in a Web browser to load more quickly by storing them locally in a cache, rather than retrieving them again directly from the Internet: similar to cache, but at a much grander scale. Also, serves as a middleman between the requesting end user and Internet while allowing administrative control, security, and caching services to the company/corporation.

public key   One part of a two-part cryptographic key that can be used to verify the owner's Digital Signature and are embedded in Digital Certificates.

RAID (Redundant Array of Inexpensive Disks)   Using multiple hard disks to provide data redundancy. RAID spreads data across several hard drives to provide fault tolerance incase of a disk crash. There are several levels of RAID. The most common are levels 0, 1, and 5.

RSA   A cryptosystem developed in 1977 used to create public and private keys using an algorithm consisting of two large prime numbers.

SAM (Security Accounts Manager)   A built-in Windows NT/2000 component that is used to manage the security of user accounts.

SAT (Security Access Token)   A security token that is used to allow users access to resources in a Windows environment. A token carries access rights that are associated with a users account.

SID (Security Identifier)   A unique security number that is associated with used users, groups and accounts in Windows NT or 2000 network. Access to processes that run in Windows NT/2000 require this unique SID and a token.

smart card   A small plastic card, similar in size to a credit card, used to store information via a microchip. The chip can be loaded with data and capable of storing more data than a magnetic strip. With respect to security, encryption keys are stored on this mobile device voiding the need for workstation storage.

sniffer   A program that monitors network traffic by comparing a list of MAC and IP addresses of approved devices with all LAN devices and negates the unauthorized ones. Used by network managers to detect problems and maintain system smoothness.

social engineering   Attempts to gain elicit access to systems by deceiving users or administrators. Telephoning users or operators pretending to be an authorized user typically carries out attacks at the target site.

spamming   A form of excessive, unsolicited bulk e-mail from the Internet where a sender is capable of mass e-mailing people and newsgroups from various e-mailing distribution lists. Spamming has the ability to overload and crash a system with its excessively large amount of data. It is not considered good netiquette to send spam.

spoofing   A form of forgery in an attempt to gain access as an authorized user. Forgery of an e-mail header allows an e-mail to be sent by an unauthorized user pretending to be someone else, thus causing the e-mail to appear to have originated from someone other than the actual source.

symmetric algorithm   An encryption process that uses a single key to encrypt and decrypt messages so they arrive to an intended receiver safely. Symmetric cryptography is less secure than asymmetric cryptography.

STP (Shielded Twisted Pair)   A type of copper cabling used in networks where pairs of wires are twisted around on another to extend the length that a signal can travel on the cable and reduce the interference of signals traveling on the cable.

S-RPC (Secure Remote Procedure Call)   Allows a secure way for a program in one computer to request a service from a program in another computer within the same network without the need for understanding specific network details. This synchronous operation runs on the client/server model; the requesting program is the client and the service-providing program is the server.

TB (terabyte)   1,024 gigabytes, approximately 1 million megabytes, or 1,099,551,627,776 bytes.

TCP/IP (Transmission Control Protocol/Internet Protocol)   The primary set of protocols used by the Internet and most networks. TCP/IP allows different networks and computers to communicate with one another.

transparency   Without hindrance or interference to the user.

Trojan horse   Destructive code that pretends to be harmless. Many viruses that masquerade themselves as something else are commonly referred to as Trojan horse viruses. These types of viruses do not typically replicate themselves. Instead they are used to introduce other destructive packages that do.

tunneling   Establishing a communications link through another networks infrastructure. With tunneling, private network packets and protocols are encapsulated and transmitted through the Internet to other private networks. In short, tunneling is most often implemented as a way to use the Internet as a means of connecting two private networks.

UPS (Uninterruptible Power Supply)   Provides a continuous supply of power to a computer system when a primary power source fails. A UPS can also protect a system from power sags.

URL (Uniform Resource Locator)   A URL is an address that points to a resource or another URL located on the World Wide Web. An example of a URL is *http://www.charlesriver.com/*.

UTP (Unshielded Twisted Pair)   A common type of twisted pair cable used in most networks. There are five categories of UTP that support different data transmission speeds. Unlike STP, UTP does not have a protective shielding.

virus   A virus is computer code or an application that is loaded and runs on a computer system with the intention of duplicating itself and other files with malicious intent. Most viruses are written to use up available system recourses until DoS (Denial of Service) occurs.

VPN (virtual private network)   A VPN is secure connection or *tunnel* that is established through a public network such as the Internet. Most VPN connections implement secure protocols such as Layer Two Tunneling Protocol (L2TP) can be used to create a secure tunnel where data is encrypted on the sending end and decrypted on the receiving end.

WAN (wide area network)   A WAN is typically made up of two or more LANs connected together to form a larger Network. WANs are usually spread over large areas. The Internet is a WAN.

WINS (Windows Internet Naming Service)   A Windows Networking Service that provides computer NetBios name to IP address resolution.

WWW (World Wide Web)   A system of servers on the Internet that provide support for pages and documents created with HTML and other scripting languages. You can access the WWW by using such tools and Web browsers such as Internet Explorer, FTP, Telnet, HTTP, and Netscape Navigator.

INDEX

Boldface page numbers indicate a review question about the topic.